The English Revolution and the Roots of Environmental Change

This study brings a new perspective to a pivotal debate: the causes of the English Revolution. It pinpoints the economic motives behind the opposition to the crown and shows their connection to the changing mindset and political transitions of the time. Distinctively, it identifies the radicalism of the mercantile sphere and the developing claim of "freedom of trade," the basis on which parliament challenged the king's fiscal prerogative. Freedom of trade was associated with rights of consent, which were asserted as a guarantee of economic interests and as a political principle. This informed the constitutional changes pushed through by parliament early in 1641, establishing freedom of trade by parliamentary control of the customs and giving the assembly an automatic place at the centre of affairs, the first requirement of representative government. Crucially, it was not the crown but parliament that appropriated the state interest, through an independent definition of national priorities. As England coalesced into a political and commercial unit, the open and communal patterns of medieval times were overlaid. The land itself came to be perceived and used in a different way. Freedom of trade had an agrarian aspect. An extended class of gentry and yeomanry occupied consolidated farms, displacing the smallholders from the common lands. With intensified marketing, the old moral restraints on trade and property died away. A more exploitative ethic undermined the balance of relationship with the land. The book makes an original connection between the English Revolution and the processes of environmental change.

George Yerby took his degree at Birkbeck College, London University, in 1986 and has since worked as a historical researcher. He specializes in the economic and political history of the sixteenth and seventeenth centuries. He is author of *People and Parliament: Representative Rights and the English Revolution*, published in 2008.

Routledge Research in Early Modern History

In the same series:

Penury into Plenty: Dearth and the Making of Knowledge in Early Modern England. **by Ayesha Mukherjee**

Violence and Emotions in Early Modern Europe, **edited by Susan Broomhall and Sarah Finn**

India in the Italian Renaissance: Visions of a Contemporary Pagan World 1300–1600, **by Meera Juncu**

The English Revolution and the Roots of Environmental Change: The Changing Concept of the Land in Early Modern England, **by George Yerby**

Forthcoming:

Honourable Intentions? Violence and Virtue in Australian and Cape Colonies, c. 1750 to 1850, **edited by Penny Russell and Nigel Worden**

The English Revolution and the Roots of Environmental Change

The Changing Concept of the Land in Early Modern England

George Yerby

Routledge
Taylor & Francis Group
LONDON AND NEW YORK

First published 2016 by Routledge

2 Park Square, Milton Park, Abingdon, Oxfordshire OX14 4RN
52 Vanderbilt Avenue, New York, NY 10017

Routledge is an imprint of the Taylor & Francis Group, an informa business

First issued in paperback 2019

Library of Congress Cataloging-in-Publication Data
Yerby, George.
The English Revolution and the roots of environmental change : the changing concept of the land in early modern England / by George Yerby.
pages cm. — (Routledge research in early modern history)
Includes bibliographical references and index.
1. Great Britain—History—Civil War, 1642–1649—Causes. 2. Great Britain—History—Civil War, 1642–1649—Economic aspects. 3. Great Britain—History—Civil War, 1642–1649—Environmental aspects. 4. Commons—Economic aspects—England—History—17th century. 5. Commons—Environmental aspects—England—History—17th century. 6. Land use—Political aspects—England—History—17th century. 7. Land use—Social aspects—England—History—17th century. 8. Social change—England—History—17th century. 9. England—Commerce—History—17th century. 10. England—Environmental conditions—History—17th century. I. Title.
DA415.Y47 2016
333.30942'09032—dc23 2015011854

ISBN: 978-1-138-93343-9 (hbk)
ISBN: 978-0-367-26385-0 (pbk)

Typeset in Sabon
by Apex CoVantage, LLC

To Cherida

who knew that the past should speak to the present

Contents

PART TWO
The Consolidation of a Political Nation

Figures

Abbreviations

Add. Ms.	Additional manuscript, British Library
Agrarian History	*The Agrarian History of England and Wales 1500–1640*, ed. J. Thirsk, (Cambridge 1967)
APC	*Acts of the Privy Council*
BL	British Library
CJ	*Commons Journal*
CSP	Calendar of State Papers
EETS	Early English Text Society
HMC	Historical Manuscripts Commission
SP	State Papers
Statutes of the Realm	*Statutes of the Realm*, ed. A. Luders et al. (London 1810–28)
The Discourse	*A Discourse of the Commonweal* (1549), ed. M. Dewar (Virginia 1969)
TNA	The National Archive
TRHS	Transactions of the Royal Historical Society
VCH	Victoria County History

Acknowledgements

I owe a debt of gratitude to those who have read and advised on this work at various stages of preparation: Colin Richmond, David Parker, Neil Davidson, Colin Mooers and Mark Levene. I must also thank Chris Reynolds for his fine photographs capturing the reappearance of the open fields at Wilstone and Hugo Butterworth for expertly illustrating the transition in architectural styles. Illustrations have a particular significance in the context of the book. I am also grateful to my editor, Max Novick, for his patience and understanding.

Introduction

The Changing of the Open, Communal Land into a National, Commercial Land, and the Neglect of Economic Effects. Is the Environment History?

In the sixteenth and seventeenth centuries, the land of England underwent a fundamental change. This had two main aspects, which were interlinked. At one level, the fabric of the physical earth was being visibly altered by a new balance of land use and new forms of occupation, more individualistic but less immediate. There were also changes in perception, and in the way the land was defined. Even as the face of the country fractured into more sharply separated holdings, it was held together by a more structured overall shape. The phrase "the land" took on a new meaning, as the body of the kingdom became fully independent, and acquired a consolidated political character.

The emerging national land created a new dimension, distanced from the actual, material land. The force of the changes on the ground was also to separate people from the underlying physical sense of the land, and to break the original closeness of that relationship. It was a critical displacement. The early modern period is known as a time of transition in many respects, but perhaps one of the most important is the least recognized, despite its urgent practical significance. Was this the moment when we began to lose our sense of the land in its real context and embarked on the divergent course that threatens such damaging consequences now, for both the earth and its occupants?

The medieval world had possessed a very distinctive concept of reality. At that time, "realism" was a belief in the objective reality of abstract notions. By the same token, concrete things could reflect universal ideas. The physical world was seen in terms of a Godly ideal, and one consequence of this was to engender a sense of respect for the land in its given character and a disinclination to interfere with that unduly. Respecting the land appeared indeed as the best way to bring forth its fruits. Respect was also entailed in the fact that agricultural labour was the one universal practical necessity. People needed to engage in a close familiarity with the land, and there was nothing to divide or distance them from it. To all intents and purposes, everyone worked on the land. Each family held a plot of land, on which their livelihoods depended. This was a context in which everyone retained an intimate sense of the reality of the land.

Throughout the Middle Ages—that long millennium of an essentially stable and coordinate civilization—the indivisible link with the land remained unbroken. In medieval times, the most common and specific meaning of "a land" was also very different from ours. It indicated the space in which precisely that closeness of relationship was enacted and fulfilled. A "land" identified the smallest discrete area of cultivation in the great common fields that surrounded each village, and it was the unit by which the equal allocation of strips in the open field was measured. It reflected the shared and direct occupation of the physical land and is a meaning that has been entirely lost to us today. In effect, a medieval "land" was defined by the working habits of the plough team; it was as much as they could conveniently cover in one journey or one day. In general, it was about a furlong, or a furrow's length, by up to four poles (or twenty-two yards) wide, depending on the character of the soil. At its greatest extent, on the drier soils which required less work and fewer drainage furrows, it was about an acre. It was essentially a unit of practice, adjusted to meet the conditions. In every respect, the medieval "land" represented an unmediated and inseparable physical connection between the village farmer and the earth.

Beyond that, encompassing all the ploughmen's lands in undifferentiated fashion, was the land in its more general being. In this dimension too, we can say that at that time the physical land could be perceived in its truest, freest form, for there were no barriers or fences between the lands of the ploughmen, and the land as a whole could stretch away quite openly and indivisibly itself. This in turn reflected the one complete physical unit on which human life depended: the single, peopled planet of the earth. The land in this sense was indeed an unbroken continuum, divided only by its own naturally varying geological character, though within it there was one line of division so distinct that vast areas of the globe appeared to be composed of water, so the land became the area that stood clear of the sea and upon which people could move and breath freely. Which brought it back full circle to the ploughman's land, which was also a land in this respect, a microcosm of the whole, as the cultivated area that stood clear of the water in the drainage furrows that ran between the lands.

In one respect, our relationship with the real land remains exactly as it was. It is still the essential basis of our existence, and always will be, for as long as that continues. This is the planet on which we are able to live and move. We still depend entirely on this same earth for our sustenance, our food, our fuel, and our shelter. Thus in what may be called an involuntary sense, the connection remains indivisible. Yet in almost every other respect, in the way that we use and perceive the land, the relationship has become tenuous to the point of breaking. By the exploitation of the earth we may soon have rendered it incapable of providing for our vital needs.

There are contradictions in our modern concept of reality, which purports to focus on the scientific analysis of the nature of physical things. In this sense we "know" much more about the land, yet we use the knowledge not to respect it as itself but rather to alter and adapt it to serve our immediate, ungoverned appetites. The balance of relationship has been lost. There is no retained understanding of the inviolable nature of the land and of our dependence upon it. Our approach to the land has become imperialistic. We rule it from a distance, manipulate its free character, and redirect its natural resources to our own convenience. Very few of us now work on the land. Even fewer own a significant portion of it. Yet the pressure on the land has increased beyond measure. We pollute its atmosphere in the remorseless pursuit of pleasure and profit, and we plunder it to the core. But as we lose the sense of what the land itself requires, we destroy its long-term sustainability, and so we come to undermine the basis of our own survival.

Our assertion of an arbitrary power over the land depends partly on the development of technologies that make it possible to induce the maximum immediate product from the earth and extract its deepest resources. But it also derives from a new concept of property rights that emerged in England in the sixteenth century. In the old, feudal dispensation, even the greatest landholders had their lands on trust, conditional on the performance of obligations owed not just to a secular lord but ultimately to God for decreed common purposes, as was consistent with a respect for the land in its given nature. By the turn of the seventeenth century, this idea of property as "conditional" was giving way to the notion that it should be regarded as "absolute". The gentry were establishing that they held their estates with complete private command. The process of enclosing the commons and open fields as individual property accelerated noticeably from the late fifteenth century onward and was sign and symbol of a greater inclination and capacity to control the land in every sense.

The landed gentry no longer wield quite such extensive power. As the towns have sprawled out into the countryside, many of us who might once have been just landless labourers have now acquired our own little piece of absolute property, though it is really just the land on which our house is built and is usually so small a patch that we think of ourselves as owning the house rather than the land. Our "land" is always smaller than the ancient working unit of the plough, and although we own it more completely than the medieval ploughman, we are not in the same intimate relationship of holding a piece of land as the basis of our livelihood. The wider ownership of property does not alter the fact that, unlike in the medieval period, landholding in the modern age is an exclusive, minority occupation.

In fact, our society of mini-property owners relates less to the real land and more closely to a different, collective usage of the word. We no longer individually work "a land" like the medieval ploughman. Rather we all *live in a land*—a political unit incorporating millions of other workers of all kinds. This takes the form of a national territory, set apart from other such lands and defined essentially by the exercise of a particular form of law and government, to which those within give their entire loyalty and adherence, as in "land of hope and glory".

At the end of the fifteenth century, and the beginning of our story, the concept of a land in the latter sense, as a complete and autonomous political unit, had not yet come into existence. In this political aspect, the medieval usage of "the land" had a looser application, for the general identification of regions. It usually indicated an area belonging to a particular race of people or a particular ruler. Thus Richard II of England could refer to "my land of Ireland". These structures of power retained a direct connection with the physical land, for the king held his realm as the overall landlord. The monarchy was the main focus of unity, but it was a blurred and shifting focus. Other independent sectional interests would exist within the realm, while the monarch might claim other realms as his. In addition, all kings were subject to the moral and political authority of the universal church. It is well said that "in the fifteenth century there was no conception in England of a unitary state: regnum and sacerdotium owed allegiance to king and pope respectively".[1]

Thus, people had not yet come to define their collective character in terms of a discrete national-political entity. Their governing loyalties were at once more local and more universal, as encapsulated in the balance of relationship between on the one hand the parish—the economic life of which was often contained within a radius of a few miles—and on the other hand the universal church, to which all parishes were directly connected and which encompassed literally everything. Society was conceived as a "chain of incomplete corporations", within the unifying corporation of the Ecclesia. The only complete singularity was the body of Christ, which embraced the whole of Christendom.[2] At this time, then, the word "state" was not used in the sense of nation state as we use it today, but in the sense of "condition"—as in the economic condition of the people or their place among the orders of medieval society. Similarly, the word "nation" itself did not yet describe a political unit. Rather, it referred to a race of people. To speak of the Cornish nation carried the same import as to speak of the English nation. On the other side of the coin, Luther could appeal to "the German Nation" as a broad race common to a wide variety of political units, independent towns and principalities. The concept of the fully sovereign nation state had not yet come into existence.

That development would take place in England during the sixteenth century. It would happen for a combination of reasons and in a variety of forms, religious, political and economic. In one important, though little recognized dimension, it was based on the emergence of a range of public imperatives, definable without necessary reference to the interests of the crown and relating rather to a series of special national priorities and characteristics to be asserted against other realms. By the end of that century, members of parliament and political theorists would be speaking routinely of the state as the unit of collective integrity and security, in just the same way as we use the term today. The events of Elizabethan times tended to consolidate the idea of England as an integrated, independent political whole, and in their wake

Shakespeare began to use the word "land" with that implication. England became "this little world . . . the envy of less happier lands . . . this dear, dear land . . . bound in with the triumphant sea".[3]

In fact, the sixteenth century would bear witness to great changes in the idea and definition of the land in all the various forms that have been mentioned above—in the way that it was owned, in the way that it was worked and controlled and in the way that it was conceived overall. The main burden of this study is to ask to what extent these developments were related. There is no doubt that in this period the physical environment, that is to say the manner in which the land was occupied and used and the character of the buildings that adorned the face of the land, altered dramatically and fundamentally, as it had not done for many hundreds of years past and would not do again for some three centuries to come. How far did these changes in usage contribute to the emergence of the idea of a politically autonomous nation state? How did the world of the ploughman's lands turn into a national land? And how far did this departure undermine our sense of the physical land as the given reality?

These are questions that have not been much addressed by the modern generation of political historians, who have tended to take little account of "the land" in either of the preceding definitions. In recent decades, historians of the early modern period have largely discounted the causative force of fundamental economic circumstances or of our developing political and national psychologies. This may seem surprising since the circumstances of the economy and the nation state are now clearly displayed as the most important conditioning factors of our lives.

The concept of the nation has come to represent the area of our exclusive loyalty and attachment. It structures and delimits our identity in a much more total fashion than did the political provisions of medieval times. In the earlier period, people owed personal allegiance to their lord and devotional allegiance to the universal church. Unlike today, there was no built-in obligation to an internalized, self-contained set of economic, cultural, legal and political arrangements and assumptions. In modern times, the nation state prefixes a complete identity for us, which separates us from all other states around and defines our mode of being, in the way that we may once have been defined simply by being human and living on the earth.

Given the constraints that the nation places on our freedom of identity, it may seem surprising that we take national independence as the definition of our "liberty". Democratic institutions can make a reasonable reality of that assumption. But the nation state also has a much more malevolent face. It acts as a concentration of subjective violence at many levels. The First and Second World Wars, with their attendant horrors and atrocities, were at root the consequence of competing nationalisms. The European Union was formed to arrest the disastrous escalation of national rivalries. In England, nevertheless, we appear to be particularly anxious to preserve our national status against dilution in Europe. This stance is not entirely unexpected, for England was the first nation state to form itself out of the previous manifestation of European unity—the universal polity of medieval Christendom. The English led the way in carving out a little world of their own and are the most reluctant to let go of it.

In view of the defining significance of the concept of nation in our lives, it might be supposed that the genesis of the form would be of great interest to historians and that the early appearance of a nation state in sixteenth- and early seventeenth-century England would be closely analysed for its causes and its effects. Yet it scarcely figures in recent historiography. Some technical aspects of unification are recognized, like the incorporation of Wales into the English state in the 1530s. But there is little acknowledgement of the rise of national sentiment and of the important role this played in the political transitions of the times. Psychology has become an important category for historians, but this tends to take an individualistic form for both subject and analyst. Radical historians find comfort in the mantra that "the personal is political", and there is truth in that. But it should not deflect us from the more direct analysis of crucial, formative public psychologies, such as the development of national consciousness and representative rights in early modern England, which have been much neglected of late.

The investigation of these developments was indeed almost precluded by the recent obsession with defining the Civil War in terms of an imagined "British Problem", which overlooked the real significance of emerging national forms in England, and diverted attention to a suppose need for the unification of the territorial components of the British Isles. This has coloured much of the writing on Civil War causation in the last two decades.[4] But it was essentially a contrivance of revisionist historians seeking alternative explanations that concealed the real causative force of the radical movements in English political life and society.[5] The Stuarts certainly had an English problem, for the parliaments in their southern kingdom were persistently seeking to enhance the representative element in the polity at the expense of the royal prerogative. In every field the Stuarts were more concerned with maintaining the powers of hierarchy than with creating unity. There was no British Problem. Regarding Ireland, it was the Irish who had the problem, facing the predatory imperialism of the English crown. Scotland neither possessed nor posed a problem and was much the most easily governed of the two Stuart kingdoms, until Charles I and William Laud provoked a rebellion by trying to impose their personal view of Episcopal order on a Kirk that was already a model of order and no threat to the stability of the faith—it merely fell short of the high ceremonialist prejudices of the Caroline regime. Charles was not aiming to bring Scotland into a new kind of consolidated state but rather to restore a traditional notion of hierarchy. One of the real problems of the Stuart kings was that they did not actually possess a developed concept of the state. It was the parliamentarians who embraced that concept, and they saw it, quite specifically, as an English state. This was an important aspect of the strength and motivation of the parliamentary side in the English Revolution.

A central point of contention about history teaching has been between conservatively inclined historians, who have favoured a focus on national traditions, and radical historians, who dislike the idea of a history of national and imperial power and prefer to think in terms of social diversity. The desire to avoid the glorification of nation and empire is understandable, for many crimes have been committed in their names. Yet they ought not to be excluded from our scheme of study. They have, in plain fact, been among the most powerful determining forces in history, and they must be recognized as such. The need is for a critical understanding of their effects, whatever they may be. And not all national traditions are to be deplored. They can involve the emergence of community values that are genuinely inclusive.

The prejudice against recognizing tradition has been reinforced by the dominant methodological assumptions of recent historical practice, the techniques of which have worked to deny the possibility of a constructive understanding of the past. The restrictive focus was put in place by a "revisionist" school of historians who, from the 1970s, set out to undermine the idea of long-term developments as described in Whig and Marxist interpretations. For this purpose, the revisionists proposed an inappropriate and impracticable standard of "empirical" evidence, asserting that history must be studied as static, factual information, disregarding the possibility of change or movement. The leading revisionist G.R. Elton produced the basic text of the doctrine that historians should take care to look at the past "in its own right" and not in terms of eventualities.[6] Despite the supposition of authenticity, this contradicted the most basic conditions of the historical discipline. It is the very essence of chronological form to move forward in time, with consequences of one kind or another; and the past can never actually be seen "as it was"—it is always seen from the perspective of the present. Revisionist/empiricism was a falsified methodology, masking a deeply political intent, and must be recognized as such if history is to recover a true balance.

In general it has been left to outside observers and logicians to expose the basic fault line in historical empiricism. Thus, E.H. Carr, in contention with Elton, declared it "a preposterous fallacy" for historians to proceed as if they were objective instruments recording discrete factual entities.[7] In more recent, post-modernist terms, Roland Barthes described historical empiricism as a sleight of hand, "in which the referent is aimed for as something external to the discourse, without it ever being possible to attain it outside the discourse".[8] In simple truth, history is by definition incapable of replicating the requirements of empirical practice, which are observation and experiment. To perform such an exercise in respect of the past would, as A.J. Ayer said, "require of the observer that he should occupy a position which ex hypothesi he does not".[9] The only point at which history can be observed in empirical form is as the present.

Notwithstanding these fundamental inadequacies, empiricism has continued to exert an undue influence on academic history. It held certain tactical advantages. To a generation that takes reference from the principles of science, the notion of looking at the past "as it was" could sound irreproachable, when baldly stated, and it had a theoretical appeal for historians seeking a straightforward, scientific modus operandi for their profession. Furthermore, it could be made to appear authentic. Because the revisionist interpretation was a negative preconception (a predetermined view that things had *not* happened), it was easier to sustain the impression that it was not a preconception at all. And although it was impossible to actually see the past "in its own right", there was a means by which that effect could be created, by the simple expedient of refusing to acknowledge any development that seemed to prefigure the future, however well evidenced it might be. Many modern historians came, in effect, to believe that they could see the past "as it was" just by ignoring any signs of what was to come. Some would assert, in tones of utter seriousness, that to note a resemblance between what came before and what came after was to fail to see the past "as it was". Revisionist/empiricism had contrived an artificial but nonetheless rigid barrier to the perception of change. In so doing, it had fostered a specific process of concealment, whereby the most powerful and lively evidences of the past were hidden from our view.

Being a misconstrued definition of the subject, it proved contradictory in its own terms. Exaggerated empirical practice tends towards obfuscation, rather than the promised clarity. The logical force of cause and effect is discarded. The strongest, clearest general evidences are obscured. Historians pursue a search for negatives, a hunt among extraneous minutiae that takes them further and further away from any coherent understanding. The exclusive focus on isolated pockets of "factual" information merely increases the scope of subjectivity. There is a proliferation of unrelated mini-theories, supposedly empirical but actually nothing of the kind. The exercise of reasoned association and the development of a balanced corroborating context are set aside. History dissolves into an ever deepening state of confusion, beyond any useful application.

The true nature of history is continuous movement and connection. There is no aspect of the past that can be isolated and looked at "in its own right". Where, then, can we turn for a formula that recognizes rather than contradicts this limitation? A convincing definition of the dynamics of the past can be found in the pages of the best known philosopher of the modern age, though he is in some ways an unlikely ally. Jacques Derrida is known for giving pre-eminence to text rather than to context in respect of the power of language. To this end, he devised the concept of "the trace", whereby no self-presence or centred meaning is ever formed because everything always contains the trace of the other. Derrida wants to tell us that we can never locate a stable point of truth. But the idea of "the trace" acquires substantial and significant effect when taken out of the realm of linguistic theory and placed in the context of real historical time. As Derrida says, "[N]o historical epoch can ever be defined simply within itself, it always contains the trace of what precedes and follows it".[10] This rings entirely true as a description of the way that history moves. How indeed could it be otherwise? And the same rule of incompletion also applies to contemporaneous phenomena. No field of activity ever takes place simply within itself; it is always in part the reflection or reaction of what moves around it. Historical phenomena operate in a continuum, in time, place, action and conception. In this study therefore we take the view that the past is best elucidated not through isolated particulars but in the fullest context of connection and corroboration. The force of things is measured not by the weight they carry in themselves but by the way they fit with others. So we need not be quite as pessimistic as Derrida. We may hope to reach a stage where our associations come to comprise a context that is sufficiently broad, clear and consistent to be regarded as a persuasive and meaningful analysis.

The fallacy of trying to disconnect and segregate the past has also acted to deny the historical influence of the land in our other definition, as an economic environment. The physical and economic aspects of our lives represent the most fundamental and inescapable realities of our existence. These are the basic conditions of our being and surviving. Yet many modern historians have managed to convince themselves that these matters have no causative effect on our history, even while the environmental crisis seems to be demonstrating the opposite.

Ironically, the sixteenth and seventeenth centuries once represented the most conspicuous example of the relevance of social and economic factors on the course of historical events. The connection was made very clearly by observers at the time. James Harrington, writing in the mid-seventeenth century, thought that the parliamentary challenge to the crown, with the ensuing Civil War, could be attributed to the change in the balance of economic weight that had taken place in the preceding century. He identified a transfer of property power from the crown and the nobility to "the people", by which he meant the middling sort of gentry, the yeoman farmers and the mercantile and trading classes. Harrington was not alone in this assessment. Christopher Hill noted that the same view was taken by all "the best thinkers among his predecessors and contemporaries". These included Sir Walter Raleigh, Sir Francis Bacon, James I, John Selden, Sir Henry Wotton, Sir Thomas Wilson, Spelman Bishop, Francis Quarles, Godfrey Goodman, Henry Parker, Henry Ireton, Gerard Winstanley and Thomas Hobbes. "It rose almost to the status of an orthodoxy".[11] To that list we can add Lucy Hutchinson, who explained the Civil War in the same terms. "When the nobility shrank into empty names the crown lost its supporters . . . when the full body of the people came rolling in upon it . . . the interest of the people, which had been many years growing, made an extraordinary progress in the reign of Henry VIII, who returning the vast revenues of the church into the body of the people, cast the balance clear on their side".[12]

The same basic theme was followed and embellished some three centuries later by Richard Tawney and Laurence Stone, who developed the idea of the rise of a commercially inclined gentry and the decline of the aristocracy.[13] Christopher Hill gave this a geo-economic perspective, postulating an economically advanced parliamentarian south and east, as against an economically "backward" Royalist north and west.[14] The economic and geographical division of allegiance was, again, generally recognized by contemporaries. Observers from right across the political spectrum, from Lucy Hutchinson to Edward Hyde, John Corbet, Richard Baxter and Margaret, Duchess of Newcastle, all concluded that parliament drew its strongest support from the ports, the commercial towns and the cloth-making districts, mainly in the south and east.[15] Underlining this as a strategic judgement, the Royalist commander Sir Nicholas Byron fully understood that the king's position was strongest and most defensible in the north and west but not sustainable in the south and east.[16]

The relevance of geo-economic factors seemed undeniable from contemporary sources, yet the most powerful tendency among recent historians has been to contradict it. Some assessments of Civil War causation, far from recognizing economic conditions as the most inescapable elements of our existence, have proceeded as if the broad socio-economic sphere did not exist at all.[17] The same kind of imbalance afflicts a recent study that defies the realities of social and political development and produces an elitist fantasy of the Civil War as "a noble revolt".[18] Other historians have noted the strength of economic change and the correlations between local economies and parliamentarian allegiance but then arbitrarily dismissed these factors as background rather than active.[19] Even historians who might be regarded as more open to considering the possibility of popular motivation in history nevertheless tend to opt for "cultural" causes, skirting round the more evident force of the economic aspect. And some go on to deprive the word "ecology" of its independent physical dimension, defining it in terms of social and religious cultures, and the opposite of economic.[20]

The "north and west versus south and east" model was, of course, an overgeneralization because there were important centres of parliamentarian support in the ports and cloth-producing areas of Yorkshire and the West Country. But the supposition of a geo-economic divide remains valid, and it is simplistic to seek to undermine this on the basis that "Royalist and Parliamentarian commitment existed in all parts of the country in 1642".[21] This is a failure to observe a proper balance of analysis, as well as an example of how a restrictive focus can be deployed to subvert the general force of the evidence. The fact that there were both Royalist and Parliamentarians in most places does not preclude the possibility that there were significantly greater concentrations in some parts than in others. This indeed is the more objective and judicious view taken by an historian of the earlier generation, Alfred Wood. He suggested that in raising the royal standard at Nottingham, the king was taking a central position between those areas that he could control easily and those that he would have to reconquer. "Though there were adherents to both sides in every shire, it is broadly true to say that the war was one between the north-western and the south-eastern halves of England".[22]

Clearly, that earlier generation of historians found no problem with including an economic or geographical perspective in their explanations. Why has the most recent generation been so determined to avoid it? Why have so many historians of early modern England chosen to put themselves in the position of discounting the causative force of the most fundamental and necessary aspects of our existence? The reluctance to see the past in a connected light probably stems in large part from the fear of applying a critical view of the way that our lifestyle has developed. This in turn has coloured the modern attitude to the most powerful critic of capitalist society—Karl Marx. Much of the historiography of recent decades has been distorted by an overreaction to Marxist history.

Marx recognized that economic circumstances were the truly unavoidable conditions of human existence. In this he was simply stating a fact, which we might regard indeed as the only absolute fact of life. More controversially, he projected a scheme of how this would work out as a process of history—advancing through a series of revolutions whereby control of the means of production would pass from the forces of feudalism, to the bourgeoisie, to the proletariat. In the decades of the mid twentieth century, the balance of intellectual opinion tended to respect Marxism as a comprehensible response to the socio-economic inequalities of the age. So although historians like Tawney and Wood were not theoretical Marxists, they saw no problem in recognizing aspects of history, such as the force of social and economic factors, which happened to reflect the Marxist view.

This balance of sentiment survived in general terms through to the 1960s. In fact, 1968 may be regarded as the moment when, for various reasons, the intellectual tide turned against Marxism and against the economic interpretation of history. The revolutionary protests of that year were led by students who had little enough to do with "labour" in any form. Furthermore, there had arisen a consumerist working class that appeared to be less the victim of capitalism than its willing accomplice. In philosophy, the so-called linguistic turn was redefining language as an autonomous structure, inwardly dictating its own terms and no longer to be employed as an instrument for identifying external truths or realities. In this postmodernist view, history came to be seen as a matter of discourse rather than cause and effect. Even leftward-inclined intellectuals like Michel Foucault began to seek alternatives to economic factors as the vehicle of historical explanation.[23]

At the same time, the crushing of the Prague Spring made the Marxist-inspired regimes of Eastern Europe appear incorrigibly authoritarian, and it became more difficult to excuse these failings by reference to the underlying egalitarian principles of the revolution. Historians who were not leftward inclined acquired a deep-set anti-Marxist phobia. With an uncompromising dislike of the direction of Marxist theory, they felt compelled to reject the movement in all its varied applications. Evidence of economic causation was rejected in advance and dismissed as a preconception. Like the empirical illusion, this created a pattern of concealment. Lecturers would assert that the idea of the parliamentarian allegiance of the middling-sort was a theory produced by Christopher Hill, though it was actually the considered opinion of almost all informed commentators at the time of the Civil War itself. Similarly, influential textbooks would suggest that to seek for explanations among the trends in religious culture could be regarded as empirical, while to observe the prominence of economic connections was somehow "reductionist".[24] The distortions of empiricist history expose its true purpose and once again obscure the real evidences of the past. Marxist history, like Whig history, was a spontaneous response to actual circumstances. They did not deny the movement of the past. Only revisionist/empiricism did that.

There thus arose the contradiction of neo-right historians heralding the triumph of capitalism in our own time, while attempting to ignore the emergence of capitalist forms in the early modern period because it reflected Marxist theory. In this consummate display of muddled thinking, historians forfeited an understanding of the most vital and inescapable economic aspects of our existence. This is not to say that socio-economic history is no longer attempted. In fact, there are signs of an emerging recognition of a need to make up for the negligence of the last few decades. There is good recent work in the field of economics and popular politics.[25] But the tendency of such studies is to be confined to a local or "micro" perspective, which limits their force and places the focus on special circumstances rather than on general balances. This may actually take us further away from the kind of coordinated assessment that might provide a meaningful analysis of the historical effects of the socio-economic field. Direct linkages between economics and general politics are still avoided. We struggle to recover that essential awareness of the connection between economic affairs and the overall thrust of our political history, though the need is urgent.

In a surreal, practical demonstration of the truths that so many historians deny, economic forces and capacities dominate our lives at every level today. Subjectively, they are taken as the principal measure of performance and worth. By them we judge not only our individual security and success but also the status and viability of the collective political unit to which we belong. Economic factors determine the way we live and the way we vote. In fact, we believe that to be tenable and justified, our society must be continually growing economically. This is regarded as the basic condition of national survival. Economic success will bring international prestige to the smallest state, while economic failure will bring down empires.

But at present, the motive force of economic activity is little more than the individual appetite for wealth and power. There is minimal exercise of rational restraint for more objective purposes. In these circumstances, our money-driven lifestyle and our insistence on justification by continual economic growth are undermining the sustainability of the physical environment and threatening the long-term viability of the earth, which is the basis of our survival. It has become apparent that, in the overall perspective, our version of economy is neither efficient nor appropriate. We are learning that the impact of the economic sphere on our history could hardly be greater and that the present course of our economic lives may soon bring historical time to an end.

This study asks whether history can suggest specific explanations for the self-destructive course on which we are set. In recent decades, historians have shown little capacity or inclination to reach a constructive awareness of the problem. Many have displayed an inability to open their minds to economic causation or to recognize the need to assess the past in terms of its consequences. We are thereby deprived of the most natural means of explaining our situation.

But these pages trace a fundamental change in the character of economic activity and our relationship to the land and ask whether there was an alternative approach, a different set of economic ethics and a more genuine kind of rationality that would have put us in a better balance with the given realities of the physical world. We seem to be abandoning the context of our natural environment. We are no longer part of what we are part. Was it in the early modern period that this divorce was set in train?

NOTES

1. J. Guy, *Tudor England*, (Oxford 1990), p. 352.
2. M.J. Wilks, *The Problem of Sovereignty in the Later Middle Ages*, (Cambridge 1963), p. 30.
3. William Shakespeare, *Richard II*, act II, scene 1.
4. For instance, D, Scott, *Politics and War in the Three Stuart Kingdoms*, (Basingstoke 2004); A. Macinnes, *The British Revolution*, (Basingstoke 2005).
5. Like other such diversions, this was set up initially in the work of Conrad Russell, "The British Problem and the English Civil War", *History* 72, (1987), pp. 395–415.
6. G.R. Elton, *The Practice of History*, (London 1970).
7. E.H. Carr, *What Is History?*, (Harmondsworth 1987), p. 12.
8. R. Barthes, *Post-structuralism and the Question of History*, (Cambridge 1987), p. 3.
9. A.J. Ayer, *Language, Truth and Logic*, (Harmondsworth 1971), pp. 25, 135.
10. J. Derrida, *Positions*, trans. A. Bass, (Chicago 1981), pp. 57–8.
11. C. Hill, "James Harrington and the People", *Puritanism and Revolution*, (London 1968), pp. 291–3.
12. L. Hutchinson, *Memoirs of the Life of Colonel Hutchinson*, ed. J. Hutchinson, (London 1846), p. 75.
13. R.H. Tawney, "The Rise of the Gentry 1558–1640", *EHR 11*, (1941), pp 1–38; L. Stone, *The Crisis of the Aristocracy 1558–1641*, (Oxford 1965).
14. C. Hill, *The Century of Revolution*, (London 1961), pp. 121–2.
15. A convenient survey is in D. Underdown, *Pride's Purge*, (Oxford 1971), pp. 10–5.

16. HMC Portland, II, p. 713.
17. C. Russell, *The Causes of the English Civil War*, (Oxford 1990).
18. J. Adamson, *A Noble Revolt*, (London 2007).
19. D. Hirst, *Authority and Conflict*, (London 1986), pp. 24–5; A. Fletcher, *A County Community in Peace and War*, (London 1973), p. 21.
20. David Underdown [in *Revel, Riot and Rebellion*, (Oxford 1985)] attributes allegiance patterns to differences in cultural and religious lifestyles and avoids the more obvious connection with economic factors; Andy Wood [in "Beyond Post-revisionism", *Historical Journal 40*, (1997), pp. 26–7] supposes a contrast between economic and ecological perspectives.
21. B. Coward, *The Stuart Age*, (London 1994), p. 205.
22. A.C. Wood, *Nottinghamshire in the Civil War*, (Oxford 1937), p. 20.
23. M. Poster, *Foucault, Marxism and History*, (Oxford 1984). Foucault came to see the relationship between "discourse" and power as the instrument of domination and the driving force of history.
24. David Underdown's stated reason for bypassing the significance of the cloth-manufacturing centres in *Revel, Riot and Rebellion*.
25. Some examples are Jane Whittle, *The Development of Agrarian Capitalism*, (Oxford 2000); Steve Hindle. "Protest and Persuasion in the Caddington Common Enclosure Dispute", *Past and Present 158*, (1998); Andy Wood, *Riot, Rebellion and Popular Politics*, (Basingstoke 2002); John Gurney, *Brave Community*, (Manchester 2007); Richard Hoyle, ed. *Custom, Improvement and the Landscape in Early Modern Britain*, (Farnham 2011); Robert Allen, *Enclosure and the Yeoman*, (Oxford 1992); John Bohstedt, *The Politics of Provision*, (Farnham 2010).

Part One

The Close of the Universal World of Medieval England

1 The Light Touch of Communalism on the Land, and the Openness of the Medieval World

Significant change was taking place in every field of activity from the early sixteenth century onward, including most fundamentally in the balance of agrarian practice. The development was in many ways slow and piecemeal, but already by the second half of the century it had reached a stage where it could be perceived as a visible departure from what had gone before and even a reversal of the traditional medieval relationship with the land.

The conceptual world of medieval times was quite distinctive, though this has been somewhat obscured by the tendency of recent historians to downplay the idea of transition between medieval and modern. The differences were important. An essential feature of the belief systems of the medieval period was the subordination of the individual, in sharp contrast to our modern perspective. At that time, society was "a single corporate entity . . . [T]he individuals comprising it play only a secondary role . . . [T]he denial of the individual in favour of the corporate reality of society is a commonplace of medieval thought".[1] It was a world indeed that defined reality not in terms of solid, singular entities but in terms of universal ideas, above all the idea of God. "The individual's relationship with society is transposed to the higher plane of man's relationship with God, and it is inconceivable that the individual should claim rights and privileges of his own at the expense of society".[2] This typical mindset had a substantial economic manifestation. It was generally held that although God had created the earth for the benefit of man, he had given this favour conditionally, only to be used for his intended purpose, which was the common good. Most commentators therefore believed that the original ideal was for the land to be held in common.[3] It was in this spirit that the medieval age was distinguished by a structure of communalism in the way that the land was occupied and worked.

This produced working relationships that were different from those of the modern age. The change is encapsulated in the fact that the most general and specific meaning for the word "land" at that time was in a sense the opposite of our current usage. It was not employed mainly to indicate the ownership of broad acres, as it is today. To the medieval farmer, the most frequent use of the word was to identify the smallest discrete stretch of cultivated ground in the open fields. It described a unit of activity, a workable area of ploughing. The length of a land, usually a furlong, was the most that a team could reasonably plough in one journey, though ploughing in as long a strip as possible reduced the need for the time-consuming exercise of turning the teams. The width of a land would be up to four poles, or twenty-two yards, and was normally determined by another practical consideration—how often a dividing furrow would need to be left for drainage. A land was also a convenient size for dividing up the village fields among the farmers, so that each got a fair share of the varying qualities of the ground. Thus the enumeration of the resulting units was the means of maintaining the disposition of the strip holdings. Each farmer would be regarded not as possessing an area of land but rather as occupying a certain number of lands, widely dispersed across the great open fields that typified the pattern of farming in England and elsewhere in the Middle Ages.

The open field arrangement was not universal, but it was the most general and is rightly regarded as characteristic of the period. W.G. Hoskins said that at the beginning of the reign of Henry VIII, "England was still medieval in every important respect . . . [A]round possibly half the villages of England stretched the hedge-less open fields that had hardly changed, except in detail for hundreds of years".[4] The open fields were common to most of Northern Europe in medieval times. In England, they were the prevalent pattern of agricultural holding in a wide swathe of country down the centre of the kingdom, from Yorkshire to Nottinghamshire and Lincolnshire, stretching broadly across the South Midlands and the Thames Valley, and down into Hampshire, Dorset and Sussex. They were less usual in the pastoral areas of the west, or in East Anglia, and in the Weald of Sussex and Kent. In fact, the extent of the open field system roughly coincided with the principal corn-growing areas of the kingdom. In other words, it was the typical structure of grain production, the crucial, core function of agriculture in medieval England.

Thus the farmers of the medieval world for many centuries elected to perform their most vital function of corn growing in a communal system. The hedge-less fields presented a different picture from our own characteristic landscape, and they were subject to a different form of tenure. Land was not individually owned in the modern manner but held and occupied in common by the farmers of the manor or the village. Each village would usually have three vast, open arable fields, each extending to several hundred acres. The farm cottages would be situated closely together in a group, forming the village. Each farmer would have a large number of strips widely distributed across the open fields and intermingled with those of other villagers. R.H. Davis gives a good description, which helps to recreate the scene. In a typical holding of thirty acres, each farmer would have ten acres in each field. But it would not be in one block. It would be ten one-acre strips scattered about the field: "some on the better land, some on the worse, some among the first to be ploughed, and some among the last".[5] To complete the impression of an interwoven tapestry, the strips were often S-shaped. A contemporary commentator captured the effect perfectly. Each farmer "had his lands not all in one gobbet of every field, but interlaced with his neighbours' lands".[6]

Under this arrangement, physical divisions were minimal to non-existent. It is likely that there were not even grass pathways between the strips. Some historians have thought that the areas marked as "baulks" may have served this purpose. But the Orwins argued convincingly that they were just occasional areas of rough, spare ground, and their only positive function was to serve as areas of impromptu pasture.[7] It was of the very nature of the open fields that the farmers worked side by side, in undifferentiated fashion, with no physical barriers between them. And the term "land's end" designated not the toe of England as it does now but the area of ground at the end of the lands, where the plough teams turned to go back or to be transferred onto a neighbour's strip.

Each strip was thus divided from the next only by the line of one of the furrows created by the plough. The system is often described as "ridge and furrow" cultivation. This pattern was produced naturally by the form and action of the plough. Each farmer began by ploughing an entire length or furlong down the centre of his strip. The mould-board of the plough (probably devised originally for covering the seed) turned a slice of soil towards the centre of the strip as it went along. The farmer then ploughed back along the other side, laying a slice of soil against the centre from the opposite direction, and a ridge was created. The farmer then continued ploughing up one side, and down the other, widening the ridge, to a width of up to twenty-two yards. The area thus ploughed formed the "land". This was essentially the area to be cultivated, as raised for that purpose above the draining water furrows on both sides. The width of a "land" varied according to the character of the ground and the nature of the soil. On particularly wet soils, more frequent furrows would be needed to maintain good drainage. In the absence of structural borders, the delineation of the strips was determined purely by the undulations of the ground and by the aim of running the drainage furrows down the slope.

These dispositions meant that the medieval farming landscape was also distinctive in being very difficult to represent pictorially. In the earliest surveys, the distribution of strips was identified not so much on the basis of defined physical entities but according to whose strips lay next to whose and by the number of lands that they contained. "Then to begin at the east side next to the highway, the parson has two lands, the lord three lands, L.B. one land, E.G. two lands, the prior two lands, the parson one land, R.X. two lands, the prior three lands, the lord two lands, L.H. one land".[8] These arrangements were governed by the communal memory of the farmers, which was usually the only record available. The result does not really enable us to visualize the disposition of holdings in the landscape. On the contrary, it underlines how the strips in each holding were widely scattered across the common fields, militating against any notion of consolidated individual proprietorship. It is sometimes supposed that, to add to the complexity, the strips were redistributed from year to year, though in fact a set of strips in the arable fields tended to remain in the occupation of one farming household for long periods. There was also some degree of consolidation and amalgamation of holdings over the course of time. This quickened markedly in the period under discussion.

But for most of the medieval millennium, the only thing that changed in the open fields was the rotation. Usually, one field would be devoted to winter wheat, one would be sown with a spring crop like barley, and the third would be left fallow. The system required careful coordination. The times of sowing and cropping and of folding animals on the stubble after harvest had to be synchronized. This was not just a matter of cooperative organization. Open field husbandry rested on a choreographed set of relationships, which again we might find hard to comprehend. Each farmer would usually be working alongside his neighbours, at the same task and at the same time, with no formal or structural divisions between. Mark Overton phrases it well: "They ploughed together, sowed together, cut hay together, harvested together, shared the tending of animals".[9] There were many advantages in the fact that they worked their strips independently, yet side by side. Each had the incentive of personal responsibility, yet the potential for coordinated effort. Proximity and the lack of structural barriers facilitated the sharing of vital resources like oxen and the pooling of labour for particular tasks. There would also be provision for mutual assistance, whereby each would contribute to a communal store, such as a stock of hay to feed the village flock in the depths of winter. The collective management of the flock was crucial. A most important aspect and definition of the open fields was the capacity to fold the village flock freely to forage the stubble of the wheat field or to manure and weed the fallow.

There was a political dimension to openness. There were important matters to decide communally, like the rotation and the drainage arrangements on the arable and the allocation of the open pasture. There were village meetings for these purposes. The system was "governed by a majority of villagers . . . and a multitude of bye-laws, laid down at a democratic village meeting".[10] It seems strange to think of a form of economic democracy functioning in feudal times. But the farming process could not be as conceptually controlled and centrally directed as it is today. The village farmers had to be relied upon for a customary working knowledge of the fields. Where there was a manorial structure, the reeve was usually elected from among the villagers and acted as much in their interests as in those of the lord.

Beyond the great arable fields lay the other crucial facility of the commons—the pastures, woods and wastes, in which each household would also possess a customary share. In a sense, these were even more open than the arable. Historians sometimes emphasize that the commons were not entirely without bounds and demarcations. But in fact the form that these took only increased the distinctive sense of openness. The use of the pastures and meadows tended to be reapportioned every year, and the borders between them share were if anything less clearly delineated than in the open field strips. On the commons, it has been well said, the divisions between the allotted areas were more "ideal" than real. They were just "imaginary lines".[11] They were usually no more than trodden out between a natural feature or stone on one side of the pasture and an equivalent marker on the other. Here indeed we may see a practical reflection of the defining notion of the medieval world that the reality was the common or universal idea, not the individual particular.

But the advantages of common rights in the use of the woods and wastes were not illusory—they were extremely valuable. There was not merely the vital facility of gathering fuel. It has been calculated that just one cow, given the benefit of rough grazing on the commons, could generate an income equivalent to half the yearly wage of a labourer, and some commoners would keep several. They could also raise lean young cattle very cheaply and sell them on to a grazier. The gleaning of wheat could keep a family supplied with flour for several months, and the gleaning of straw could provide the raw material for productive crafts like straw platting. These facilities might sometimes be free for all but were sometimes communally allocated according to a scale. This system of shared rights in the open fields and commons, medieval communalism, was the basis of the livelihoods of the people of the villages for hundreds of years. It had great social and economic strengths, although it was clearly a way of occupying and using the land that was qualitatively different from the landholding arrangements to which we have since become used, where consolidated farms or properties are individually owned and are firmly and formally fenced off from those of the neighbours.

The question of what inspired the communal system of farming is of great interest but difficult to answer with precision. It seems to have emerged in Anglo-Saxon times, as did the use of the heavy plough, which benefited from it. The open fields are first mentioned in Wessex and the Midlands in the seventh century and had reached Yorkshire and Lincolnshire by the ninth.[12] Their spread coincided with an extended period of settlement and clearance, which lasted from the seventh until the thirteenth century. Debate has centred on whether communal practices were the product of principle or convenience, but they probably owed something to both. Not enough account is taken of the fact that the process of clearance was communally driven. Although some may have been done under the direction of interested landlords, by far the greater part was the initiative of villagers joining forces to extend their scope for cultivation and grazing. This entailed a fair distribution of rewards. As ground was cleared, each farmer was naturally allocated an equal share in the fields. For poor farming people, there were great advantages in sustaining cooperative relationships. It enabled them to share the heavy plough and teams of oxen, which were beyond the means of any single peasant household. And leaving the holdings unbounded made it easier to move these facilities from one to the other.

Above all, it needs to be recognized that farming in common was appropriate to the conceptual inclinations of the time. The open field system appears to have been specific to the medieval period. It fitted the distinctive character of that era, as it would have done in no other. It became the natural corollary of an age when land was thought of as being "held" conditionally, as a trust, rather than "owned". Conditions were imposed at every level. Under the feudal system, the king was regarded as the overall possessor of the land, in one sense by original conquest but in another upon condition that it was employed, as God desired, in the general interest. The king granted lands to his principal lieutenants, conditional on continued services. They in turn granted lands to their tenants, also on condition of services and payments. Fealty and mutual obligation held society together. Everybody owed services. Nobody owned land in the complete sense that we would recognize today. Autonomous rights, whether of "states" or of individuals, were not the guiding light of the Middle Ages. The maintenance of communal balances took precedence over the assertion of private interests. Agrarian communalism was the grounded reflection of a philosophy that subordinated the individual to the corporate being and the particular to the universal idea. "The individual's relationship with society is transposed to the higher plane of man's relationship with God, and it is clearly inconceivable that the individual should be in a position to claim rights and privileges of his own at the expense of society".[13]

The religious dimension is the most difficult to elaborate, though it may be of great importance. The establishment of the open fields and commons predated the feudal system, probably by several hundred years, but was in much closer correlation with the career of the medieval church. Communal farming appeared at roughly the same time as the extension of the authority of the universal church of Christendom. It is not easy to demonstrate but very easy to envisage that medieval farmers would have cleared and cultivated the land in the light of the religious principles that they took as their essential guide and guarantee. There is little doubt that the processes of agrarian communalism received specific encouragement from the influence of the church, in a society where the clerical estate laid down rules for every kind of activity. It was certainly supposed by most commentators that, according to the Bible, the earth had originally been held in common and that this "natural and blessed community" should still be regarded as the ideal arrangement.[14] For centuries to come, the medieval world remained a church-led society that sought more than any other to apply the precepts of the faith to social and economic conduct.

A sense of Christian mutualism can be found in the poets of the age. Chaucer's picture of the Ploughman captures the distinctive balance of priorities.

> A true swinkere and a good was he,
> living in peace and perfect charity.
> God loved he best with his whole herte
> At all times, though him gamed or smerte.
> And thanne his neighbour right as himself.

He would threshe, and thereto dyke and delve
For Christ's sake for every poore wight
Withouten hire, if it lay in his might.[15]

Piers the Ploughman too would "swink and sweat and sow" for others as well as for himself. In a phrase that brings the medieval agrarian world to life, he declared that he was "following the plough for poor men's sake". "I'll give them food myself, as long as the harvest doesn't fail . . . bear ye one another's burdens and so ye shall fulfil the law of Christ".[16]

The universalism of medieval life is encapsulated in the poets' evocation of the overall religious context and in W. E. Tate's description of the divisions between holdings on the common lands as more "ideal" than real. The philosophical or psychological tendency of the Middle Ages was to conceive of things in terms of heavenly or universal exemplars rather than grounded specifics.[17] So it was a time when spiritual or moral forces seemed just as "real" as anything else. In this sense too there was less awareness of a solid or autonomous individual dimension to things, whether as subject or object, and there was less inclination to separate things out than became the rule in modern times. Morally and physically, as well as economically, the medieval world was dedicated to the maintenance of connections and balances. It was not simply a case of giving active force to the principles set down in religion. All things were integrally associated. "There is no divorce of temporal from spiritual . . . since all distinctions are lost in the unity of the universal society".[18] So the physical was held to have spiritual content, and the spiritual was believed to have physical effects. As we shall see ahead, the support of spiritual forces was continually invoked to give practical assistance to the work of the farmers in the fields. The spiritual, the intellectual and the physical could be regarded as operating in the same plane. All was conceived in a continuum. The communal use of the open fields was a function of that same universality.

This, then, was the distinctive mindset that accompanied the life of the open fields. It was very different and in many ways completely opposite to the arrangements that obtain in the modern world. The full force of the distinction becomes apparent if we think of it not just as a form of tenure but in terms of the way people fitted into their environment. The whole context of the agricultural scene was *open*, not only in the absence of hard and fast divisions between holdings and the cooperative forms of agrarian organization, but also in the way that people related to the land itself. Their presence in the fields was of a notably different kind than we might recognize today. Communalism had a distinctive physical character. Each morning, the farmers and their animals would leave the village in a long procession and fan out into the fields. They were there by no one's particular right of possession but rather as a working community, all performing the same tasks at the same time. Their situation was dictated not by the direction of employers but rather through a shared physical activity. This reflected relationships that were more equal and less controlling, both between the farmers and between them and the land. It was not just that the forms of ownership were less consolidated but also that the human imprint was less specific, hard and permanent than it later became. Medieval farmers occupied the land not by the structured forms of hedges and fences, secured individual properties, detached farmhouses and technical command, but simply by their own collective presence, just by being there, out in the great open fields, together.

This picture of the lighter, yet closer touch upon the land is not just a logical impression. It can be illustrated. Some of the ancient, open fields around the village of Wilstone in Hertfordshire (where the author's ancestors once held common rights) disappeared under reservoirs after enclosure. But in a recent drought, the reservoirs dried out, and the lines of the open field strips resurfaced (see Figure 1.1). This offered a rare sight of the actual strip holdings and illuminated the character of the medieval agrarian world as just described. It is possible to visualize the farmers working side by side along their "lands", and in the contrast between this communal context and the structured modern landscape around it, we can gather the distinction between a countryside occupied just by a collective working presence and one occupied by the fixed barriers of property and fence. Here and at Broughton, the hard line of the hedges gives a powerful sense of the open scene being cut off by enclosure (Figure 1.2).

It is a distinction that was recognized in the sixteenth century itself. It was understood and enunciated by the MP Henry Jackman, in 1597. Ironically, Jackman was envisaging the end of the open fields. He was speaking in favour of the practice of enclosing and consolidating the common holdings for pasture, defending it against the familiar accusations that it led to depopulation and dearth. He pointed out that this threat had been at least as prevalent in the medieval period, before significant enclosure had taken place—"when the whole land lay in open fields, subject only to the hand of the husbandman".[19] This provides a fine definition of the different, communal way of relating to the land. It was a relationship that was at once freer and more direct, an unmediated connection, when the fields were "subject only to the hand of the husbandman" and not yet tied down by division or brought under individual ownership.

So the farmers inhabited the land just by their collective presence. Often the commons would be full of the village farmers and their families, going about their customary tasks and occupying the fields in the course of their work. Here indeed was Piers' "plain full of people".[20] But conversely, at times when the work had for the moment ceased and the procession had reformed and returned to the village, or in the months when the sowing and the weeding had been done, and the great wheat-sown field was at its height, the land must have appeared quite empty, at least in terms of human activity. There were no other structured signs of occupation, and the fields were in effect left to themselves. Then, the only distinguishable features of the countryside were the varying undulations of the land and of the earth in its different phases. Only the changing forms and colours of the crops on the ground betrayed the fact that the hand of the husbandman had been there, reflecting the distinctive balance of relationship between the farmer and the land.

The same can be said for the restrained low profile of the village in the landscape. We can picture a rural scene where tiny, low-slung stone—or more usually timber-built—village cottages were huddled together in the middle of vast open fields and commons. These settlements made only a minimal mark on the landscape. Many of the villages must indeed have been completely inconspicuous. This was, however, of less importance to them than it might be to us today because almost all of their economic activities were contained within their own cottages, fields and commons, and they could be content that their community was visible to God. It was therefore quite fitting that as far as human strangers were concerned, the presence of the village would be vouchsafed only if the church possessed a prominent spire. This too represented the universal idea of which they were all part.

The one underlying link with the modern age was that the farmers were working on the same basic geological resources as we do now, but in every other sense it was a very different world. They ploughed the same earth, but they worked on a set of assumptions that we no longer recognize. And over the course of time, that other world disappeared. The clustered medieval village dispersed and sometimes vanished altogether, as the commons and the open fields were enclosed, and the buildings of the consolidated farms or estates were set up more prominently at a greater distance from the original village site.

There thus emerged the very different scene of the pinned-down, parcelled-up and closely controlled landscape that we are used to looking at today. From the window of the train or the car, we gaze out on the picture that has characterized the English countryside since the end of the seventeenth century—the familiar patchwork of hedged-off fields and the authoritative shape of individual farm properties, laid out around the solid core of the brick-built farmhouse and its outbuildings. In this world, the grip upon the land became at once much tighter and yet somehow more detached. Our relationship with the land became less immediate. We rarely see more than one farmer in a landscape now, who is usually hidden in the cabin of a vehicle, yet the mark of humankind on the land is everywhere and indelible.

We can picture the medieval scene and describe and even illustrate its distinctive character, but can we ever get an actual sense of what it was like, when that world has long since disappeared and ours is so different? There are a few places where the context has lived on in sufficiently complete form to give us a real understanding. Chailey Common in East Sussex still exists in something like its original use (Figures 1.3 and 1.4). For many centuries, the local cottagers depended on the common for grazing their stock and for cutting wood and bracken for fuel. But by the twentieth century, this traditional economic use had ended, and the birches and the bracken threatened to take over and turn it into a recreational wood, like most modern-day commons. Chailey Common was, however, to recover its original shape. It was taken under protection, with a view to preserving its working character. Sheep were reintroduced to graze the pasture and to keep the birch saplings and the brambles at bay, and parts of the common were regularly mown to manage the bracken. Thus the balance of resource and working activity, of interdependence and natural control, has been sustained. This too can be illustrated. The surviving Chailey Common presents a varied, yet open and coordinated landscape, and it is possible to visualize the equilibrium that held good for so many centuries—the equilibrium among the village farmers, between them and the land and indeed within the land itself.

In respect of the open field system, there is just one village in which it has survived almost complete. In Laxton, Nottinghamshire, just east of Sherwood, the open fields still exist in working agricultural use (Figures 1.5 and 1.6). The three great fields are much reduced in size to less than a third of their original extent, and they are worked not in an independent communal context but as an appendage to various individual farms in the vicinity. Nevertheless, they are still farmed much as they would have been in medieval times, and it is surprisingly possible to get a sense of the distinctive character of the world of the open fields. The Orwins, surveying the Laxton fields in terms of the pattern of strip divisions, were struck by the diversity of size, shape and direction of the holdings, which they thought looked somewhat arbitrary and inconvenient, a rather jumbled affair, determined only by the contours of the land.[21] But this is perhaps to look at them too much in a modern perspective. We need to think of them less in terms of shapes on a map and more in terms of a time when shapes on a map hardly existed and when the essential relationship was between the body of the land and the hand of the husbandman.

At Laxton, it is still possible to experience the sense of openness. The unmediated perspective survives. Despite the fact that the arable fields are much reduced in size, the visitor can still stand at the southern edge of West Field and look across it as a vast open landscape, stretching away to the skyline, uninterrupted by any fixed mark of human tenure. The farmer said that when it comprises the winter-sown field and everyone sows wheat, it looks at its height like an ocean of corn. And since it was once three times its present size, it is easy to conceive that it must indeed have appeared as an entire universe, unending and indivisible. At the time of viewing, it was the spring-sown field and therefore under a variety of crops, as selected by each farmer, as is permissible at this phase of the common field sequence. Rather unexpectedly, the sense of openness and coherence is not lost. The farmer said it was a good time to view the common system in the West Field, since the strips were at that moment differentiated by being under different crops or at different stages of cultivation. He was bailing the corn on one of his strips. His neighbour on one side had grown beans instead, while his neighbour on the other had not yet cut his corn. In each case, the only division was the line where one crop met the other, and the only pathway was the bare ground where each crop tapered out. An impression of incoherence might have been expected, but on the contrary, a deep sense of peacefulness prevailed. There was no rude interruption of fence or hedge, and the land was running free. The only distinctions were those made by the hand of the husbandman and the character of the crop. Here indeed was the equilibrium of the communal way, which held the people in an even balance and left the land at liberty. And when you have gathered in the sense of this, it stays with you as a kind of fundamental freedom, a generosity of nature, open and unbroken. Then it is the later imposed, enclosed landscape that comes to look fractured and jumbled.

This was the contrast that John Clare perceived, as the hedges divided up the wild moors of his childhood.

> In little parcels, little minds to please.
> With men and flocks imprisoned, ill at ease.[22]

Figure 1.1 The exposed open fields of Wilstone. (Courtesy of Chris Reynolds.)

Figure 1.2 The open field ridges at Broughton. (Courtesy of Chris Reynolds.)

Figure 1.3 The surviving commons at Chailey. (Photo by author.)

Figure 1.4 The surviving commons at Chailey. (Photo by author.)

Figure 1.5 The surviving open fields at Laxton. (Photo by author.)

Figure 1.6 The surviving open fields at Laxton. (Photo by author.)

NOTES

1. M. Wilks, *The Problem of Sovereignty in the Middle Ages*, (Cambridge 1963), pp. 24–5.
2. Ibid.
3. John Ap-Robert, *The Younger Brother His Apology*, (St. Omer 1618), p. 7.
4. W. G. Hoskins, *The Age of Plunder*, (London 1976), p. 1.
5. R. H. Davis, *A History of Medieval Europe*, (London 1970), p. 197.
6. *A Discourse of the Commonweal*, (1549), ed. M. Dewar (Virginia 1969), p. 121.
7. C. S. and C. S. Orwin, *The Open Fields*, (Oxford 1954), pp. 45–8.
8. Fitzherbert, *Surveying*, (1523), in *Certain Ancient Tracts*, ed. R. Vansittart, (London 1767), p. 70, BL 1508/1096.
9. M. Overton, *The Agricultural Revolution in England 1500–1850*, (Cambridge 1996), p. 24.
10. W. G. Hoskins, *Age of Plunder*, p. 12.
11. W. E. Tate, *The English Village Community and the Enclosure Movement*, (London 1967), p. 32.
12. F. Stenton, *Anglo-Saxon England*, (Oxford 1947), p. 277.
13. M. J. Wilks, *The Problem of Sovereignty in the Later Middle Ages*, p. 25.
14. John Ap-Robert, *The Younger Brother His Apology*, p. 7.
15. Geoffrey Chaucer, *The Canterbury Tales*, ed. F. N. Robinson, (London 1957), p. 22.
16. William Langland, *Piers the Ploughman*, ed. J. Goodridge, (Harmondsworth 1966), pp. 81, 84, 87.
17. See J. H. Huizinga, *The Waning of the Middle Ages*, (Harmondsworth 1976), chap. 16.
18. M. Wilks, *The Problem of Sovereignty in the Later Middle Ages*, p. 29.
19. J. E. Neale, *Elizabeth I and Her Parliaments*, (London 1953), II, p. 342.
20. Langland, *Piers the Ploughman*, pp. 25–6.
21. C. S and C. S. Orwin, *The Open Fields*, pp. 96–7.
22. John Clare, *The Mores*, in *John Clare: Major Works*, (Oxford 1994), pp. 167–9.

2 The Dismissal of the Saints and the Disappearance of the Universal Church

In a sense, the medieval agrarian context was by definition an ideal world, an agricultural system with moral purpose and direction, where economic life reflected a Godly idea. In practice too, medieval communalism seemed to create a series of perfect balances among the farmers and between them and the land. But in terms of the basic need for subsistence, those relationships left something to be desired. We may eulogize the open equilibrium where the farmers occupied the fields by their working presence rather than by possessive command, but it was in one respect a sign of weakness. There was a downside to the "lighter touch"—it betrayed the fact that the farmers' control of the land was not sufficiently tight and effective to ensure the long-term health and survival of their families and communities. Medieval agriculture was not capable of sustaining the kind of indefinite population rise that became usual in the modern age. An increase in numbers would inevitably hit a ceiling of limited productivity, and a demographic setback would result.

It is clear that in the early fourteenth century, in the decades before the Black Death, there was an agricultural crisis due to overpopulation, soil exhaustion, a shortage of manure and a lack of underdrainage that compounded periods of bad weather. Consequently, "large and growing sections of the population had been reduced to a condition in which they could keep body and soul together only in years of moderately good harvests".[1] In bad years the death rate rose, and it took more than a century for farming productivity to recover a beneficial balance.

This is not, however, to contradict our eulogy so far as to say that the open field system was incapable of increased productivity. In fact, there is a current tendency to suggest otherwise. Robert Allen has written that the common fields were quite able to adopt new methods, and although enclosure was more likely to lead to investment and improvement, there was surprisingly little difference in yields, according to comparisons made in 1800, when the agricultural revolution of the early modern period had taken effect.[2] He calculates that on average, the surviving open fields had attained 86% of the increase in yields achieved on the enclosed farms since 1500.[3] But the conclusion that the open fields were in general terms equally efficient is probably in need of qualification. Enclosed systems were naturally innovative, while open field culture, which relied heavily on the observance of custom, was not inherently conducive to new ideas. And it is probable that in the sixteenth century, when improvement was at a premium, enclosed systems did lead the way. If the open fields were approaching parity by 1800, it may be because in the course of three centuries they had ample opportunity to adopt not only the techniques but also the general approach of their enclosed neighbours.

So although we can reject the opinion once held that communal farming was "impervious" to change, it should be clear that in its original context open field husbandry was a characteristic element in a medieval system of agriculture, which was very limited in its methods and in its ability to improve them. The average grain yield of medieval times was about ten to twelve bushels an acre, which was rather less than half what was being achieved by 1800, when the "agricultural revolution" had taken place. The medieval approach has been described as "wholly un-progressive".[4] It is true that the Middle Ages saw no significant advance in farming technology after the introduction of the heavy plough, the iron share and stronger harnesses, which are all credited to Anglo-Saxon times. There was a gap of many centuries before the age of agricultural improvement began in the sixteenth century, and some historians would even defer it to the seventeenth.

Yet in a sense it is wrong to expect the medieval world to have produced many changes in agricultural method. It was an age that did not really possess a concept of technical improvement. The essential aim of that society was not to advance but to maintain a pattern of God-given balances. Some of the failings that are usually mentioned may be regarded as interesting reflections of the general tendency to treat everything in the perspective of an undifferentiated whole. There seems to have been no concept of seed selection, and the practice of broadcasting seed was very inefficient, in a way that could ill be afforded in terms of productivity. Dung and marl were the only recognized fertilizers, and there was little attempt to adapt them to specific soils or uses. All was as one in the Godly society. God gave to everything its nature, and it was to some degree inappropriate to make changes or distinctions in the dispositions that God had made.

It was, then, a reflection of both the conceptual basis of medieval society and the precarious state of medieval agriculture that the farmers placed a great deal of faith and reliance on invoking spiritual assistance to bring about the meteorological and elemental conditions necessary for a successful harvest. In some respects, they were seeking to influence matters that would always be beyond the power of farmers to arrange. But at another level their recruitment of the supernatural betrayed their inability to improve their lot by more practical solutions and in some degree a disinclination to seek them. In this sense, the farmers' practical devotions were a function of the "lighter touch" and the absence of a tighter control over the land. And once again it resulted in a form of activity where they related to the land simply by *being there*. It has been suggested above that the village farmers who filled the open fields and commons occupied the land more by their collective working presence than by commanding it as property. The spiritual dimension—the very distinctive pattern of communal and devotional mobility with which they moved around the countryside—was also part of that relationship. Piers' "plain full of people" naturally included the pilgrims, priests and itinerant friars. The villagers existed in a remarkable revolving network of pilgrimage, procession and communal supplication for day-to-day spiritual aid with their endeavours in the fields.

Most prominent among the villagers' external devotions, and most vulnerable to the increasingly vociferous condemnation of religious Reformers, was the intensive veneration of the Saints and their relics for the physical and agricultural support systems that they were thought to provide. The Reformers regarded these assumptions as false imaginings, which involved the worshipping of Saints in a manner that should have been reserved for God. On this view, it was legitimate for the villagers to ask God to bless their endeavours in the fields but "superstitious" to give worship to particular Saints for particular purposes. The village farmers, for their part, were simply embracing what appeared to them as the most efficient means of applying their devotions to their most vital practical needs. There were considerable advantages in addressing their requests to the Saints rather than direct to the Almighty, precisely because the Saints could be regarded as task specific. You could seek aid from "St. Appolonia for the toothache, St. Blaise for the throat, St. Legeard for the eyes, St. Loy for the horses, St. Anthony for the hogs".[5] At Bury, there were relics over which you could give supplication "for avoidance of weeds growing in corn".[6] The accounts of the Bishop of Durham in the early fifteenth century contain a payment for signing sixteen cattle with St. Wilfred's signet to protect them from the murrain.[7]

There is no doubt that these externalist practices were still being followed with enthusiasm on the eve of the Reformation, as is pointed out by Catholic-inclined historians like Scarisbrick and Duffy.[8] In this study too we have found much to appreciate in the ways of the medieval village and would by no means regard it as a society in decline and ready to be superseded. But what the Catholic view seems not to recognize is that the very popularity of externalist observances could lay them open to attack by more sceptical minds. J.H. Hale has noted this potent tension in the context of Europe generally: "All the traditional aids to Catholic devotion were enthusiastically taken advantage of; indeed it was in response to the vigour of these devotions that satire at the expense of superstitions and the externality of religious observances was at its height in the late fifteenth and early sixteenth centuries".[9] So when Erasmus visited one of the most popular pilgrimage venues in England—Our Lady at Walsingham—he "went there to scoff".[10] The shrine was the centre of some extreme supernatural beliefs, such as the conviction that the chapel had arrived direct by magical flight from the Holy Land. Yet Erasmus was also obliged to note that it was "the most frequented place throughout all England, nor can you easily find in that island a man who reckons on prosperity unless he yearly salutes [the Virgin] with some small offering".

It was significant that in England, the theological aspect of Protestantism was less to the fore than elsewhere in Europe and that, partly for this reason, the condemnation of externalist devotions was an especially important and central element in the Reformist critique from the beginning, long before Protestant beliefs as such had gained a substantial foothold. The attack on externals was launched with vigour in the early 1530s, under the generalship of Thomas Cromwell and Thomas Cranmer, as soon as Henry's banishment of the papacy had put the old church on the defensive. It is important to recognize that this was not just an iconoclastic bulldozing of the functions of the Catholic Church. The Reformers were putting forward a well conceived alternative to outward devotions and indeed to the medieval mindset as such. The traditional externalist practices were characterized without hesitation as sorcery and delusion. The English *Primer* of 1535 set the tone: "Some people have been deluded of long time about the veneration of Saints and such like things".[11] So when Cromwell's agents arrived at the shrine of Derfel Garden in Llanderfel and found hundreds of pilgrims at the site, offering supplication to the great wooden image for the health of their cattle, it was taken not as a sign that the practice should be sustained as a focus of popular devotion, but rather as confirmation that it should be suppressed as a cause of widespread public delusion and a considerable waste of money.[12]

The attractions that these systems of belief held for the people of the villages are very comprehensible. For the farmers, they offered a source of hope that there was some possibility of influencing the all-powerful physical and environmental forces that governed their work and their livelihoods. For the church itself, the attraction was also partly pragmatic. The concept of the physical powers of "Saints and such like things" gave the church a hold on popular affections that it could probably never have achieved if its services had been confined to the theological exposition of the afterlife. Although there were always some churchmen who were uncomfortable with the excessive influence of externalist practice, the general tendency was to accept the psychological and financial power that it brought to the church.

It would be wrong, however, to assume that these practices were an obsession restricted to the common people. Kings were equally likely to indulge in such devotions, though probably for different purposes. Indeed, as already suggested, we should not underestimate the extent to which the externalist approach was inherent in the medieval mindset at every level. Johan Huizinga has come closest to elucidating this. "If the medieval mind wants to know the nature or the reason of a thing, it neither looks into it to analyse its structure, nor behind it to enquire into its origin, but looks up to heaven where it exists as an idea . . . to reduce it to its universal principle . . . the mind is not in search of realities but of models, examples, norms".

Consequently the ideal is reflected back in the form of magical substance: "the tendency to ascribe a sort of substantiality to abstract concepts . . . it cannot be denied that the medieval mind frequently yielded to the inclination to pass from pure idealism to a sort of magic ideal, in which the abstract tends to become concrete". There was thus "an ultra-realistic conception of everything that related to the Saints", who became "familiar and life-like".[13] This was the basis on which physical and spiritual forces could be conceived as working together in a continuum.

So the village calendar was full of Saints' days, at which there would be feasting in acknowledgement of aid received or anticipated. By the early sixteenth century, there were more than fifty holy days to be celebrated during the year, and the work of the farms proceeded in close conjunction with the ceremonies of religious supplication. The farming year opened at the feast of Epiphany, which was the last act of the Christmas season, when Christ was made manifest to mankind, represented by the Magi. Frankincense and myrrh were diffused over the fields as a blessing, marking the farmers' own recognition of the divine birth. With the universal presence of Christ thus affirmed, work could begin, but first there were the rituals of the plough. On Plough Sunday, the first after Twelfth Night, the ploughs were blessed. Across the great Midland Plain, the heart of arable country, plough lights were kept burning in many churches. Often a plough was brought into the church and a light placed before it, so that the beneficial force could be conveyed direct to the object. The next day, Plough Monday, was celebrated with feastings and collections to keep the plough lights shining, and a plough was carried ceremoniously through the village streets. The sanctification of the plough was the most graphic illustration of the assumed interdependence of the practical and the spiritual. But Reformers were unconvinced and complained of "the conjuring of ploughs".[14]

The movable feasts of Easter coincided with the celebrations of spring, the triumph of new life over death. The sixth Sunday in Lent was Palm Sunday, commemorating Christ's entry into Jerusalem. The consecrated host was carried in procession, with the laity bearing branches blessed by the priest. The practical essence of the matter was the welcoming of the coming season. "In the overwhelming majority of English parishes, Palm Sunday was almost wholly a festival for the blessing of foliage, a Christian rite of spring".[15] The 25th March, or Lady Day, was dedicated to the Annunciation of the Blessed Virgin. It was also a date when rents and tenures were resettled, coming at a break in the farming cycle, when many of the fields lay fallow.

May Day was associated with the feasts of Saint Philip and Saint James, though it was at heart an ancient fertility festival, whose dominant characteristics were flowers, maypoles and dancing as communal foreplay. This inaugurated a continuous celebration of summer, with games and church ales continuing through the warm months to July, all in celebration of the season, with the added purpose of raising money in support of the farming community and the parish. Usually later in May, leading up to Ascension Day, came Rogation week. This was the occasion for the perambulation of the parish bounds, but it was also a continuation of springtime revelling and license, with feasting, dancing and games, and was sometimes labelled "roguing" week on that account.[16] The more legitimate sense of the word "rogation" came from the Latin word to "ask", so the procession was not just a marking of the boundaries but also a collective supplication of the community for spiritual protection for the crops now growing in the fields. It was a presentation, in particular and in total, of all the endeavours for which they hoped to receive the aid and approval of spiritual forces and to be blessed with success. There would be open-air services at which the land and boundary markers would be sanctified. Banners would be carried aloft, celebrating the deeds of the Saints and the life of Christ.

Midsummer Day, the 24th June, was marked by the feast of the nativity of Saint John the Baptist, but in essence it was a celebration of the summer solstice and the turning of the year. Its principal component was fire. There were torchlight marches and the guns of the Midsummer watch. There were bonfires, and burning wheels were sent rolling down hillsides. In the cider-producing areas of the West Country, the bonfires had a traditional supplicatory purpose, to "bless the apples".[17] In other ways, however, the effect sought was more direct, for the heat and flames were thought to purge the air of the infestations and infections that otherwise threatened to come to their peak at the same time as the crops. The externalist practices did not entirely preclude a "scientific" intent.

At the beginning of August came Lammas (or loaf mass), inaugurating the harvest season. Entertainments were now focused on the markets and fairs. There were harvest suppers to mark the gathering of the most vital crop in the village. Then came Harvest Home, or so it was hoped. Nothing can ever have been named with more eager anticipation or more anxiety. And nothing was celebrated with greater joy, if and when it arrived. If success had been secured, the last wagon of corn was brought in, as ever, in procession, bedecked with garlands, and mounted with ritual images made from straw gleaned in the fields.

Then the village community was moving towards Christmas, when there was a natural interval in the work of the fields, and the farmers could devote themselves to the celebration of the birth of Christ. At the same time, the streets and houses were decorated with the evergreens that served in remembrance of the growing year now past and in anticipation of the next one soon to begin.

Considering the coordinated strength of the revolving network of spiritual and agricultural association, as well as the interdependence of these most critical aspects of medieval life, it comes as rather a surprise to find that the structure collapsed in relatively short order. In a sense, its homogeneity was its weakness. It fell as it had lived, all together. In the two decades between 1529 and 1549, the core and substance of Merry England were destroyed. This tells us again of the force of the alternative that was being established. And the job was done before even the first tentative steps were taken towards defining a Protestant settlement, which confirms that the English Reformation had its own distinctive character and that it is a mistake to measure it merely in terms of when Protestantism arrived. "Protestantism had made a relatively slight impact before the reign of Elizabeth".[18] In England, theological Protestantism was very slow to mature, and some might say never really attained a full shape. Even the Elizabethan settlement was a compromise, which retained many features of the old faith. But the attack on what the Reformers and others, even including the otherwise conservative Henry VIII, could define as "superstitious" was immediate and central from 1530 onward, and it had been pushed to a conclusion before Elizabeth came to the throne. The recognition of this tells us much about the nature of the English Reformation and about the general psychological changes that were taking place.

The English *Primer* of 1535 had set the agenda: "[S]ome people have been deluded of long time about the veneration of Saints and such like things".[19] The attack came step by step but with impressive conceptual certainty and determination. The Injunctions of 1536 sharply reduced the number of holy days and abolished many Saints' feasts. Some historians have suggested, understandably, that this was in itself a severe weakening of the general force of popular devotions.[20] It was the beginning of the end. The Injunctions of 1538 launched the decisive assault on the basis of externalist practice, ordering the removal of any image that was thought to have been the subject of superstitious pilgrimage or worship. The veneration of relics was banned. The Saints who survived in the calendar were to be regarded merely in the spirit of memorial and no longer invoked as active agencies. All lights in the churches were to be extinguished, apart from those on the altar. The inseparable working relationship between the fields and the heavens was over. The lights that had so long been kept burning in supplication for the practical aid of the Saints and the benefit of the ploughs were performing that task no more.

The cult of the Saints would never recover. Some other aspects of communal ritual remained, but their principal supports were being cut away, and they were on borrowed time. Ronald Hutton sums it up. "The four months after Henry's death, January to May 1547, saw the last replay of the traditional spring and summer ceremonies of church and populace".[21] The new order of the Edwardian Protectorates was more rigorously Reformist, and the assault on externalist practices was duly pushed to a conclusion. The Injunctions of 1547 outlawed pilgrimages and even the processions that had accompanied the Sunday mass inside and outside the church, thus "striking at the heart of one of the principal expressions of medieval communal religion".[22] The destruction was ordered of all shrines and pictures of Saints, and all images to which offerings had been made. In February 1548, a proclamation forbad the major ceremonies for the blessing of candles at Candlemas, the blessing of ashes on Ash Wednesday and the blessing of foliage on Palm Sunday. It also banned Plough Monday, destroying the symbolic core of the working link between the two great imperatives of the farm and the faith.

It is interesting that the most characteristic aspects of externalist religion were not restored in the reign of Mary.[23] Even with the encouragement of Mary's brief Counter-Reformation, there was no general revival of the popular preoccupation with pilgrimages and the cult of the Saints. This was in part because the focus and framework of externalist practice had been removed when the ecclesiastical houses were demolished, but it was also because the sustained Reformist assault over the course of two decades had permanently undermined the social and psychological strength of the old practices. The dislike of "superstition" commanded a consensus beyond the ranks of committed Protestantism. The Elizabethan settlement of the church, which was to prove long-term, reflected the distinctive balance of English Reform. It was not so decidedly Protestant as to wholly repudiate the old beliefs and ceremonial. But it was the end of "outward" religion as a source of physical effect. "By 1570 the process of transforming the churches of England and Wales from settings for ritual into preaching houses was complete".[24]

The dismissal of the Saints ran concurrently with and was greatly accelerated by the dissolution of the ecclesiastical houses. The monasteries, abbeys, priories, chapels and chantries had been the most important presentational stages of the Saints' cults. The religious houses were the venues for pilgrimage, the guardians of the relics and the focus of the ritual. They were the service centres between which the revolving round of pilgrimage, procession and supplication flowed. They provided the framework within which the villagers moved collectively through the countryside. The ecclesiastical buildings were also the most prominent and significant architectural features in the landscape. They were in every sense the principal reference points of medieval society.

So there can never have been a more dramatic transformation in the life and character of the countryside than that produced by the disappearance of the ecclesiastical houses and the patterns of mobility that revolved around them. The fact that this very distinctive context is difficult for us to recreate, even imaginatively, underlines the depth of the change. We are used to hearing of the significance of the redistribution of church property, mainly into the hands of the gentry, in "the largest transference of land since Domesday".[25] But we are not sufficiently aware of the revolution that had occurred in the very tenor of rural life and its relationship with the physical environment. Medieval society was a world of communal mobility and common working patterns that were bound together by religious forms and spiritual forces, in a way that is completely alien to us today. It was a context in which every aspect of existence was part and parcel of every other, which helps to explain why it vanished almost overnight.

The agents of its demise were a laity with an overriding appetite for land, reinforcing a Reform movement that rejected the temporal effects and jurisdictions of the clerical estate, in league with a king who had decided that the universal church was a contradiction of his personal ambitions and his authority within the kingdom. It was a powerful combination, as it needed to be, in order to destroy a long-standing and pervasive socio-religious culture. The assault began in 1536 with an attack on the smaller, weaker houses. Almost four hundred of them were destroyed at the first strike. Warming to their task, in 1539 Henry and Cromwell moved on to the destruction of the larger houses. The last of them fell in March 1540. Within four years, the core structure of the church-led civilization that had lasted almost a millennium had been reduced to rubble. Then, by an act of 1545, renewed in the reign of Edward VI, the colleges, hospitals, free chapels and chantries were all suppressed. There had been no less than twenty-four hundred chantries and chapels. They were all gone by 1549.

The disappearance of the ecclesiastical houses and the appropriation of the church properties by the laity constituted the sharpest and most substantial change in the way that the land was furnished and occupied. The ubiquitous physical face of the universal church in the landscape suddenly vanished from view. It was the end not just of a form of a religion but a form of civilization. Today we view the ecclesiastical houses as picturesque ruins, haunted by the ethereal presence of their erstwhile inmates. But we need to envisage them as working communities, the focal points of an economy, and a source of social and cultural provision—reflecting in every way that inseparable connection between the physical and the spiritual that characterized medieval life.

The independent church had defined the appearance as well as the character of the medieval world. Along with the castles, which had been far fewer in number and much quicker to decay, the ecclesiastical houses had been the largest, most prominent and impressive buildings in the medieval countryside. The scale and suddenness of their destruction are underlined because, unlike the castles, the buildings and properties of the church were not in decline at the time of their downfall, at least not physically. Although there may have been some falling away in the spiritual and social welfare services that they were expected to provide, this was in contrast to their physical vigour. In fact, Henry and Cromwell caught them at their most vulnerable and luxuriant development. The enormous wealth of the clerical estate was made manifest, to both envious and admiring eyes, by the growing magnificence of the church houses. The ecclesiastical sector was much the liveliest area of building and architectural activity in the late fifteenth and early sixteenth centuries. There had been extensive enlargement and improvement of the greater houses. "The monasteries dominated the building scene down to the 1530s, either in themselves or in the parish churches they were rebuilding".[26] It was the churches and ecclesiastical houses that impressed foreign visitors.

So on the eve of the Reformation, the ecclesiastical complexion of the countryside was if anything becoming more marked than ever. There was little competition in sight. Only the religious buildings displayed what can be regarded as a fully formed architectural style. They manifested qualities of symmetry, relief and ornamentation that secular buildings had not really begun to emulate. It was the religious houses that engaged and expressed the leading public aesthetic of the time, as befitted the status of the church as the defining influence in social and political culture.

By contrast, there was in the early decades of the sixteenth century little indication of a truly distinctive secular building programme emerging. The first signs of an aspiration towards secular style were appearing in the somewhat equivocal shape of the castellated house. The great age of the manor house proper would begin in the reign of Elizabeth. In a sense indeed this development waited upon the reduction of the medieval church. And it depended not simply on the transfer of the abbey lands and wealth into the hands of the gentry but also on the withdrawal of the all-embracing patterns of spiritual recourse and communal mobility with which the old church had been associated. It entailed their replacement by a new social context, characterized by a more entrenched individualism and a different form of public unity.

More will be said on that subject in later chapters, but one of the foundations for a new kind of unity can be identified at once. In an obvious way, the break with Rome was the basis for a clearer definition of the concept of a "land". The dissolution of the monasteries and the rejection of papal authority cut away the crucial dimensions and guarantees of the universality that had characterized the medieval world. In the process, the scope for a discrete national entity occupying its own public space was created. As Henry so succinctly put it, the clergy had been discovered to be but half his subjects, and with his usual ruthless lack of ceremony, the rival focus for the loyalties of the English people had been promptly dispatched. After the Act in Restraint of Appeals of 1534, Christians in England no longer owed their ultimate allegiance to the Bishop of Rome. The English church was separated and made independent of the Pope but subjected to the king and his parliament. Then, with the demolition of the ecclesiastical houses, the living link that had existed between the monasteries and their founding institutions on the Continent was also severed. The independent, universal authority that had encompassed the whole of Christendom as one church polity had been banished from the realm. England stood for the first time entirely upon its own feet, as an autonomous nation state. Equally formative was the legislative sovereignty of king-in-parliament, which had authorized the dismissal of the Pope and which now comprised a powerful inner dynamic binding the nation together, just as it bound the people to its statutes. The surviving church was subordinate to the secular power in a way that the old church had never been. Thus the secular sphere was set free from the limitation of existing merely as a corporation within the general body of the universal church.

The core component of the sovereign secular state, and indeed of the new concept of fully sovereign statute law that had brought it into being, was the individual. The overall authority of statute law derived specifically from the idea that it was made with the consent of "everyman", and not by mere diktat. In the succeeding pages, we shall see the individual conscience beginning to assert an autonomous sphere and a right of consent in the face of clerical power. But it is useful to note the more general underlying force of the disappearance of the universal church in changing the balance of initiative between society and self-perception. The medieval world under the aegis of that church had not recognized the existence of the freestanding individual that is so familiar to us today. At that time, "society is a single corporate entity . . . the individuals comprising it play only a secondary role . . . the denial of the individual in favour of the corporate reality of society is a commonplace of medieval thought.[27] Individuality was subsumed in the body of the community. The medieval outlook turned instinctively towards the universal rule that transcended the individual. In the Middle Ages, "men disregarded the individual qualities and fine distinction of things, deliberately and of set purpose, in order always to bring them under some general principle".[28] The Church of Christendom was the unifying force. The Body of Christ was the only complete singularity. But with the expulsion of the universal church, the all-enveloping unity of Christ was deprived of its actual, practical location. The corporate ideal was dismembered, the world could be seen in the light of differentiation, and significant scope was created for the emergence of the self-possessed individual.

NOTES

1. M.M. Postan, *Medieval Economy and Society*, (London 1972), p. 34.
2. R.C. Allen, *Enclosure and the Yeoman*, (Oxford 1992), p. 15.
3. Ibid., p. 134.
4. A.L. Poole, *Domesday Book to Magna Carta*, (Oxford, 1954), p. 50.
5. *The Rationale of Ceremonial*, ed. C.S. Cobb, Alcuin Club (London 1910), pp. 49–50.
6. E. Duffy, *The Stripping of the Altars*, (Yale 1992), p. 384.
7. K. Thomas, *Religion and the Decline of Magic*, (Harmondsworth 1973), p. 29.
8. J.J. Scarisbrick, *The Reformation and the English People*, (Oxford 1984); E. Duffy, *The Stripping of the Altars*.
9. J.H. Hale, *Renaissance Europe*, (London 1971), pp. 239–40.
10. C.S.L. Davies, *Peace, Print and Protestantism 1450–1558*, (London 1984), p. 145.
11. C. Butterworth, *English Primers*, (Pennsylvania 1953), pp. 90–1.
12. G. Williams, *The Welsh Church from Conquest to Reformation*, (Cardiff 1962), pp. 495, 502.
13. J. Huizinga, *The Waning of the Middle Ages*, (Harmondsworth 1976), pp. 161, 206–7.
14. R. Hutton, *The Rise and Fall of Merry England*, (Oxford 1994), pp. 16–7.

15. Ibid., pp. 20–1.
16. D. Cressy, *Bonfires and Bells*, (Berkeley 1989), p. 23.
17. Ibid., p. 26.
18. R. O'Day, *The Debate on the English Reformation*, (London 1986), p. 152.
19. C. Butterworth, *English Primers*, pp. 90–1.
20. E. Duffy, *The Stripping of the Altars*, pp. 394–8.
21. R. Hutton, *Rise and Fall of Merry England*, p. 76.
22. E. Duffy, *The Stripping of the Altars*, p. 452.
23. R. Hutton, *Rise and Fall of Merry England*, p. 101.
24. Ibid., p. 109.
25. J. Thirsk, *Agrarian History of England and Wales*, (Cambridge 1967), IV, p. 302.
26. W. G. Hoskins, *The Age of Plunder*, (London 1976), p. 12.
27. M. J. Wilks, *The Problem of Sovereignty in the Middle Ages*, (Cambridge 1963), pp. 24–5.
28. J. Huizinga, *The Waning of the Middle Ages*, p. 207.

3 The Reordering of the Physical and Intellectual Spheres

The English Reformation was not primarily or even necessarily about the assertion of a new theology. It was most centrally about the displacement of an externalized form of worship by an internalized one. The picture of the churches transformed "from settings for ritual into preaching houses" captures the outline shape of the change. But the underlying psychological shift ran broad and deep and was reflected in many other aspects of life. The disappearance of the working Saints, the ecclesiastical houses, and the shrines, relics and images that were their stock in trade marked the passing of a world in which the faith was manifested primarily in terms of physical effect. Now the comprehension of the Bible became the one crucial feature of worship. This was not just a technical adjustment in the idea of what constituted the most efficacious route to salvation. It involved more generally a reordering of the respective status of the physical and intellectual fields.

In fact, the cause of Reform had a *rationalist* thrust that should be more clearly recognized. The Reformers rejected externalist devotions not just because they deflected the act of worship away from God but also because they were intellectually unjustifiable—they could not be rationally defended. The force of this made a crucial contribution to the success of the Reform movement, but its significance is apt to be overlooked, for two main reasons. One is that a rationalist trend can easily but mistakenly be represented as inconsistent with a religious perspective. The other is that the recent generation of historians have been inclined to ignore the emergence of rationalism in an obsessive anxiety to avoid the idea of "progressive" advance. There has been a paranoid fear of acknowledging any kind of onward movement. As already argued, this is a debilitating and self-defeating phobia that results in the concealment of the real evidence of change. To observe the fact of a development is not to "preconceive" it or to embrace it as progressive. In these pages, the balance of approach in the medieval world is not regarded as "backward" in any overall, qualitative sense. There is no assumption here that the process of internalisation is to be regarded as an "advance", or that intellectual procedures necessarily convey more meaning than physical. But we do need to be clear that this was precisely what was being assumed in the sixteenth century; otherwise we miss the essence of the psychological transition that was taking place in early modern England. At the core of it was a change in the balance of initiative between the physical and intellectual spheres.

The Reformers rejected externalist devotions specifically on the grounds that they could not be rationally supported. They were consistently characterized as delusions that denied not only the true nature of the faith but also the proper exercise of reason. The Injunctions of 1538 deplored the idea of giving credence to practices "devised by men's fantasies . . . as in wandering to pilgrimages, offering of money, candles and tapers to images or relics, or kneeling or licking the same, saying over a number of beads not understood or minded on".[1] The insistence on a faith that was "minded on" is the crucial indicator. The new demand was for a defensible, intellectual understanding of the faith as the core of the devotional process. Thomas Cranmer, who presided over the first two decades of Reform, best summed up the movement as an assertion of the primacy of the intellect. The veneration of images gave "no manner of honour", neither that due to God nor "such as is due to his reasonable creatures".[2] The Reformers deplored the popular belief that specific physical observances could produce specific physical effects. As Keith Thomas pointed out, they regarded these things as magic tricks. The Lollards called it "necromancy", and the sixteenth century Reformers usually referred to it as "sorcery".[3] Again, Johann Huizinga is the best guide to the link between idealism and magic. "The hyper-idealist mentality . . . the tendency to ascribe a sort of substantiality to abstract things . . . the medieval mind frequently yielded to the temptation to pass from pure idealism to a sort of magic ideal".[4]

The Reformers believed that externalist devotions were false, both spiritually and intellectually. What, then, *did* they regard as practices giving the honour due to God and his "reasonable creatures"? They wished in effect to focus on an aspect of the faith that could be regarded as demonstrably real and that could be engaged by the self-conscious intellect. Cranmer again clarified the dichotomy: "Purgatory, pilgrimages, prayers to saints, images, holy bread, holy water, holy days, meats, works, ceremonies and such other . . . the Word of God hath got the upper hand of them all".[5] So the comprehension of the Bible offered the route by which "his reasonable creatures" could give honour to God. The religious aspect could be compatible with the rationalist because the Bible could be regarded as *evidence* of the Word of God, there to be studied and elucidated. Ironically, it was important that the words of Scripture *were* now physically *there*, in printed and vernacular form, in a way that they had never been during the medieval period. The publication of the Great Bible in 1538 was intended specifically to open up the scriptures to the populace. It was still expected that their findings would not contradict the official view, but this limitation in itself revealed a moment of transition. It is of some significance that the rejection of the "magic ideal" and the new emphasis on the faith as an intellectual exercise were occurring *within* the continuing context of religious belief and a devotional society. The proposition that belief must be based on an open relationship between the individual and the Bible was the root from which the liberty of conscience might grow. Again, in the terms of our introductory chapter, it shows how change might come via the "trace" of one phenomenon within another.[6]

External observances were seen then as contrasted with the elucidation of scripture. Unlike the Bible, the physical powers attributed to Saints, relics and images were not susceptible to intellectual verification—they were mere suppositions, magical effects, or "delusions", as the Reformers constantly asserted. This differentiation and the dispelling of the "magic ideal" reflected a changing relationship between intellectual and physical capacities. This indeed was at the core of the new psychological balance. In both the agricultural and the religious fields, physical phenomena were being set at a greater distance, in a relationship that was more detached and less immediate. In both cases and in every sense, the physical aspect was being critically subordinated and no longer incorporated in the same conceptual plane as the intellectual and spiritual.

The changing balance was displayed in various forms. The arrival of farming literature in the sixteenth century signalled a new approach in the agricultural context, whereby practices were assessed in terms of rational improvement and increased profit, essentially divorced from moral and spiritual considerations and looking to employ the conditions of the market rather than the good offices of the Saints. The further applications of this transition in respect of the farming world will be described in chapters ahead. But the emergence of the rationalist turn is most strikingly illustrated in the spiritual field, in particular with the beginnings of a tolerationist trend, based principally on the same separation of the intellectual and physical spheres. This is a development of obvious importance, both in itself and as reflecting the changing mindset of the time, but it has been obscured by the restrictive tendency of recent historians to view any sign of a tolerationist policy in the sixteenth century as just an enforced pragmatic ploy for keeping the peace.[7]

In truth, the "trace" of tolerationist principle is very clear. The need for some movement towards liberty of conscience was strongly implied by the unusual centrality of the evangelist drive in the English Reformation. If the active reception of the Bible was paramount, then it must be opened out and the individual conscience allowed the necessary degree of freedom to respond. Previous centuries had presupposed a passive acknowledgement of the faith. The idea that beliefs could be imposed by force and eradicated by fire had been integral to the medieval balance of relationship between the physical and intellectual fields. In effect, under that regime, the mind was not accorded an independent realm of its own beyond the physical body. This was now beginning to change.

The traditional stance was first challenged in the early 1530s by a persistent attack in the House of Commons on the arbitrary nature of heresy proceedings. The early stages of the Reformation Parliament are quite understandably noted for the essential contribution made by the Commons to the destruction of the independent legislative power of the Church and its replacement by the overall supremacy of parliamentary law. This was the change that underpinned the English Reformation and indeed much else of consequence in the political developments of the succeeding century. But side by side with the assault on the church's judicial authority, though less often remarked, was a determined campaign against the character of heresy proceedings.

These Commons' initiatives were indeed intimately connected. This can be seen in the position taken by men like Christopher St. German, who had led the assault on the independent legislative jurisdiction of the church and who further believed that the clergy had no right to condemn people in law at all. This injunction naturally came to include the question of prosecution for heresy. St German held that the Gospel gave the Apostles authority to preach but not the right to prohibit others from doing so. He thought that this, like every other kind of sovereign provision, should be left to the king and his parliament.[8] Restricting the powers of the church to a purely spiritual influence was an important aspect of sketching out a sphere where the mind or conscience could be free of physical force, and St German's argument soon came to be extended.

So among the first set of complaints against the church in 1529 was a draft bill challenging the manner in which heresy trials were conducted. The first allegation was that the clergy exercised an arbitrary power to interpret any criticism of clerical behaviour as heresy, even though it might be exposing clerical vices of an essentially secular nature. The bishops and their officers "perversely and uncharitably have and do arrest and convent before them such as either preach, speak or reason against their detestable and shameful living". The accused were then kept in prison "under the colour and name of heresy" while the clerics worked up a charge against them. The church authorities produced "no proofs or witnesses in their behalf but their own uncharitable suspecting". The victims were then obliged "either to abjure and remain in perpetual prison, or else to be burned and utterly cast away for ever".[9] In these words it is already possible to gather the idea that the clergy should not be allowed an arbitrary power to define the nature of heresy, and that burning people for dissent was neither appropriate nor desirable.

The proposed remedies would have put some notably strict controls on the implementation of heresy charges and the discretion of the clergy. No one was to be arraigned for "preaching or informing anything contrary to the Catholic faith or doctrine of the church" if the clerical authorities had "any malice or grudge, variance or displeasure with the person so arrested". The ordinaries were also to be obliged to produce "two or three honest and credible" witnesses, who were also to be without malice, variance or displeasure against the accused. The identity of the accusers and the details of the charge were to be made clear, and the accused allowed bail while awaiting trial.[10] This burden of proof and the demand for due process contradicted the very nature of heresy trials. The right of the clerical authorities to simply impose their will would have been ended, and the right of accused persons to state their case affirmed.

The attack was driven home to a yet deeper level in the petition in which the Commons also launched their complaint against the clerical power of legislation as such. It seemed indeed that their most powerful specific objection to the workings of the church courts related to heresy proceedings. The petition took aim at the actual techniques of inquisition. It described how "the said ordinaries use to put to them subtle interrogatories concerning the high mysteries of the faith and are able quickly to trap a simple unlearned or yet well-witted layman without learning and bring him soon to his confusion".[11] This was to take issue with the clergy's assumption of a right to impose themselves upon the minds of others and to construe opinions as they saw fit. In effect, it was to question the basis of the concept of inquisition, which was that the individual mind did not possess freedoms and integrities of its own but could be treated as a passive or corporeal medium upon which the clergy, as the authorized possessors of the single God-given truth, were free to imprint their own interpretation. The Commons' petition marked the genesis of a new psychological balance that will be seen emerging more clearly in the pages ahead—the growing belief that the mind ought not to be treated as if it were a physical entity that the clergy could twist or force into the shape they required. The petition also carried the challenging implication that although the view of a "well-witted layman" might not chime exactly with those who were in command of "the high mysteries of the faith", it might nevertheless be valid in its own terms. This was underlined when the petition went on to protest that heresy trials left the individual with a stark choice of either denying his beliefs or dying for them. He must either abjure "and thereby lose his honesty forever" or stand by "his innocent truth" though he would face "death by fire, and so be utterly destroyed".[12] The suggestion was then that, when the accused were forced to recant, they were not necessarily renouncing an error but were instead being denied the honesty of their own instinctive understanding of the faith, or their "innocent truth". The ramifications of this were very striking. Here quite clearly was the germ of the idea that the individual mind or conscience should be allowed its own integrity, even though it might not be precisely in tune with the orthodoxy as defined by the authority of the church.

The regime of Henry VIII was relatively sparing in its use of force against "heresy", but burnings did take place. The persecution of Robert Barnes, William Jerome and Thomas Garret in July 1540 displayed all the unacceptable faces of heresy proceedings against which the Commons had protested so vehemently a decade earlier. The real transgression of the accused was political—they had provoked the anger of Stephen Gardiner by supporting Thomas Cromwell.[13] There was no trial in open court. The charge against them, implicitly Anabaptism, was wholly unsupported and, as far as can be ascertained, completely groundless. They appear to have been simply mainstream evangelists—preachers of the Gospel. They were arbitrarily declared guilty of grievous heresies and destroyed.

But in late Henrician England, the concept of destroying heresy by fire was not quite as normative as it had once been, and it did not pass without protest. In fact, the burnings provoked further declarations and developments of a tolerationist tendency. In 1542 Henry Brinkelow took an important step in that direction in his complaint to parliament about the cruelties lately practiced by the bishops. At one level he echoed the protests voiced so distinctly in the Reformation Parliament. He reiterated the critique that heresy proceedings were utterly lacking in due process of law. "Men are condemned to death and have not the witnesses come face to face openly in court". By this means "a man shall be condemned to most cruel death and not brought to open justice . . . death, death, even for trifles, so that they follow the high priest in crucifying Christ . . . we have a law and by this law he ought to die . . . this mori, die, die never went out of the priests' mouths since that time, and now they have poisoned the temporal rulers with the same". He was not only rejecting the arbitrary powers of the priests in heresy proceedings, but also revealing a profound distaste for the concept of physically destroying people for their faith.

He took this further by articulating what had been implicit in the protests of the 1530s, that is to say the idea that these questions should be debated as between equals with an equal right to the justification of their own interpretations. "What reason is it, or what Christian heart will say it is right or lawful, that men may be at a controversy of a matter of life and death, the one should be put in the hands of the other . . . it is not lawful that preachers should be put into the hands of the bishops with whom they contend". Brinkelow suggested that if the bishops were also to have their views "tried by an indifferent judge", they would not be so hasty in accusing others. He was envisaging a disputation rather than an inquisition. This was confirmed in his notion of what forms of discipline should be regarded as legitimate. If a transgressor persisted in refusing to accept the manifest truth of the Bible as demonstrated by an indifferent judge, he should be banished. If he returned, he could be deprived of life, but even in this case the sentence carried a crucial qualification. Physical punishment was justified only "because he breaketh the command of the temporal powers, and not for his faith's sake, for faith is the gift of God only".[14] Here for the first time clearly expressed was the assertion that faith could not be forced, reflecting a growing conviction that there was no place for coercion in religious affairs. The intellectual sphere was being separated out and ring-fenced against the physical.

The reality of the tolerationist trend was given striking confirmation at the centre of affairs in the policies of the first Edwardian Protectorate under the Duke of Somerset. In 1547 all the heresies laws, Lancastrian and Henrician, were repealed. It was a powerful reflection of the growing tendency to set the conscience free of physical invasion. The very title of the old act, De Haeretico Comburendo, encapsulates the previous lack of inhibitions about applying corporeal solutions to cerebral questions. The medieval world did not accord the mind a sphere of integrity separate from the body. So they had no problem with the supposition that heretical ideas could be destroyed by fire. The 1547 repeal of the heresy laws stands as an explicit repudiation of that concept and as a bold assertion that in truth the use of force was not appropriate in matters of the faith. It was supplemented by the reversal of the restrictive measures of Henry's later years, which had withdrawn the right of free Bible access from the general laity. The determination to reopen the Bible to all the people was a clear illustration of the consequences of the distinctive centrality of the evangelist imperative in the English Reformation. Somerset and his advisers appear to have believed that open access to the scriptures would bring accommodation and consensus, so the scope for this to take place must be allowed.

The movement towards denying the efficacy of force in religious affairs was expressed most powerfully in the writings of various members of Somerset's circle. His chaplain, William Turner, produced the first clearly tolerationist work. Turner broke new ground with an articulated argument that faith could not be determined physically, but only intellectually. He provides another interesting example of "the trace of what precedes and follows"—retaining a concept of heresy but rejecting the traditional means of dealing with it by force. Turner asserted that heresy, whether Catholic or Anabaptist, should not and could not be eradicated by fire. "Some would think that it were the best way to use . . . material fire and faggot. But me thinks, saying is no material thing that we must fight withal". He pointed out that heretical beliefs were ideas rather than objects, and they could be countered only by other ideas and by arguments adduced from the evidence of the Bible. "Or else we are likely to prosper but a little in our business". "When the enemy is a spirit . . . the weapons and the warriors that would kill the enemy must be spiritual".[15] Saying is no material thing. This was the essence of a changing psychological balance, by which the intellectual sphere was separated out from the physical. The contrast between material fire and spiritual fire is specific and conclusive. The conscience was being redefined as "no material thing", and death by burning could no longer be accepted as a proper way of addressing it.

The same imagery was used in *A Discourse of the Commonweal*, written at much the same time. The *Discourse* was mainly devoted to economics, with a similar approach to Fitzherbert, based on rational analysis and the acceptance of the profit motive. Religious differences were treated as a separate issue in a brief addendum but with a not dissimilar assertion of the rational. "It is not in the disputation of the truth as it is in a fight or a wrestling . . . let the matter be quietly discussed . . . that neither party do use any violence against the other . . . freely consent and determine the matter".[16] Thus, as the intellectual context was separated out from the physical world, the analytical capacities of the mind were highlighted. It was through the development of scientific faculties that physical phenomena would now come to be viewed. And it is in recognizing the contribution that the Reform movement made to the rationalist turn that we come to fully understand the crucial psychological shift of the sixteenth century.

The reign of Elizabeth gave an important degree of established acceptance to the emerging conception of the mind as an independent sphere, with rights of consent. The idea was embraced at the centre of the polity. The Queen and her Council specifically and consistently denied any right or desire to force or invade consciences. Elizabeth seems to have shared the position chalked out by William Turner and the author of the *Discourse* that the conscience was "no material thing that we can fight withal". She believed in effect that the conscience was a private world that could be subject only to the free comprehension of the Bible. In fact, she coined the phrase that summed up better than any the ring-fencing of an inviolable intellectual sphere—she declared that she would make no windows into men's souls.

In this spirit, Elizabeth resisted the proscription of Catholicism until the direct threat to her safety and the security of her realm became palpable, and even then she strove to sustain her tolerationist principles. In the face of the Northern Rebellion of 1569 and the Papal Bull of 1570 inciting her deposition, she stood steadfastly by her insistence that consciences were not to be invaded. In January 1570, she had a Declaration of the Queen's Proceedings "prepared for general dissemination, and carefully corrected the text in her own hand, emphasising its liberal doctrine by her amendments".[17] She promised that she would maintain her tolerant approach and that, as long as her subjects did not break the public peace, they would continue to enjoy her customary mildness "without any molestation, by way of examination or inquisition . . . of their opinions for their consciences in matters of faith, remitting that to the supreme and singular authority of almighty God, who is the only searcher of hearts".[18] Although she believed that lenience could be politically prudent, her position also revealed her genuine commitment to the tolerationist principle that the mind or conscience was an independent and inviolable sphere.

Accordingly, Elizabeth vetoed a parliamentary bill that would have enforced the taking of communion upon all. The measure was much debated, and one of the Queen's supporters in the Commons, the lawyer Edward Aglionby, gave further elucidation to the tolerationist position. In a striking speech, he amplified Elizabeth's sketch of a free, noncorporeal realm of experience into which windows could not be made. He rejected the idea of compulsion to communion on the basis that this was to belie the nature of the conscience, which could not be determined by force. "The conscience of man is internal, invisible and not in the power of the greatest monarch in the world . . . in no bounds to be straightened, in no bounds to be constrained".[19] It was a fine attempt at an exposition of a crucial conceptual transition. He was seeking to characterize the way in which the conscience was being redefined as physically inviolable. The difficulty of the exercise underlines the significance of the process of separating out the two spheres. The mind or conscience was now seen as an autonomous space that was "invisible". It was "internal" yet uncontainable. It had no solid outward form, so it could not be physically engaged. It could be reached only through intellectual exchange and consent. Thus the intellectual sphere was being allocated a medium of its own that could not be subject to physical determinants.

These strands of thought, like so many of the vital political and psychological changes of the sixteenth century, were brought together in the work of Richard Hooker in the 1590s and to that extent were made orthodox. In Hooker we find the clearest expression of the intellectualization of the faith and its association with a rationalist trend. It is Hooker who most fully articulates what this implied in terms of the assumption of a right of consent, which had emerged in various fields of activity since the 1530s. One of Hooker's central objectives was to challenge the subversive tendencies of militant Puritanism. But he did not proceed by simply asserting the need for unity and hierarchical obedience. He says very little about principles of order. It was to the concept of consent that he turned for his arguments. He questioned the Puritans' assumption of unique correctness. He said that this did not reflect the proper process by which truth might be established. Hooker was, above all, a rationalist. The truth was to be ascertained by reason, not by revelation. It could not be based merely on what anyone cared to claim that God's spirit had made known to him. The scriptures were to be elucidated by "some star or light of reason or learning".[20] It could not be just "what he himself imagineth that God's spirit doth reveal unto him . . . if it had come from God . . . the same God that revealeth it to them, would also give them power of confirming it to others . . . with strong and invincible remonstrances of sound Reason, such as whereby it might appear that God would have all men's judgements give place unto it".[21]

Hooker proceeded to elaborate the essential basis for determining disputes by reason, or by the exercise of a right of intellectual consent. Arguments or analyses were to be accepted only on the basis of demonstration. Only thus could one mind be expected to accept the proposition of another. "It is not required, nor can it be exacted at our hands, that we should yield unto anything other assent than such as doth answer the evidence which is to be had of that which we assent unto".[22] Here were distinct echoes of William Turner's original proposition that disputes could not be determined by force but might be resolved by argument. And Hooker was indeed in the same latitudinarian or tolerationist camp. But he was developing the exercise of reason a stage further by giving a defining actuality to the force that the rational mind might be expected to accept—that is the force of evidence. He was thereby rounding out the concept of a right of consent, in all its varied applications. And by the same token he reveals the common ground by which the reordering of the intellectual and the physical spheres could apply in the fields of both the faith and the farm. There is the same assumption of intellectual detachment and control, whereby the world comes to be analyzed as evidence, rather than as given.

The process of allocating greater independent freedom to the intellectual sphere and imposing more critical and conceptual restraints on the force of physical phenomena represented an important change in the public mindset, which was reflected in many fields of activity. There was also another way in which the internalization of the faith worked to reduce the special significance of the physical dimension in medieval society as a whole. The "external" context was being undermined both in particular and in general. The assault on externals was not just a condemnation of the corporeal or "outward" aspects of the act of worship; it was also a rejection of the concept of religious devotions carried on in the physical world *outside*. The truly distinctive feature of medieval religion and indeed of medieval society as a whole had been in assuming an area of direct reciprocal involvement between the operation of the spiritual world and the work of the land. Now the whole extraordinary revolving network of pilgrimage, procession and communal supplication in the streets and fields was being abruptly terminated. It is difficult to overestimate the significance of the change in social and environmental psychology. It has been rightly said that the patterns of conviviality became more restricted, and focused within the public house, so that by 1600 the communal sociability of the parish as a whole had been replaced by the "fragmented sociability of the alehouse."[23] The same shift had taken place in the religious sphere in respect of its relationship with the agricultural world. The sixteenth century witnessed the withdrawal into the church of the devotional life that in medieval times had been played out as part and parcel of the open countryside.

Most of all in the context of the present study, it is important to recognize what this meant for the way people related to the landscape around them. It was suggested above that the village farmers occupied the fields and commons more by their working physical presence than as property. The remarkable round of association and supplication between heaven and earth had been integral to that collective, open occupation of the countryside. This was now being brought to an end in all its aspects.

NOTES

1. *Visitation Articles and Injunctions of the Reformation*, ed. W.H. Frere and W.M. Kennedy, Alcuin Club, (London 1908–1910), pp. 33–43.
2. Thomas Cranmer, *Miscellaneous Writings and Letters of Thomas Cranmer*, ed. J.E. Cox, The Parker Society (London 1846), pp. 100–3.
3. K. Thomas, *Religion and the Decline of Magic*, (Harmondsworth 1973), pp. 58–61.

4. J. Huizinga, *The Waning of the Middle Ages*, (Harmondsworth 1976), pp. 206–7.
5. Thomas Cranmer, *Writings*, ed. J.E. Cox, pp. 350–2.
6. Ibid., pp. 6–7.
7. This assumption is prevalent, for instance, in *Tolerance and Intolerance in the European Reformation*, ed. O.P. Grell and B. Scribner, (Cambridge 1996).
8. Christopher St. German, *Answer to a Letter*, (London 1535), p. 108.
9. TNA SP2/N, f. 21.
10. Ibid., f. 22.
11. TNA SP6/7, ff. 211–20.
12. Ibid., f. 221.
13. J.J. Scarisbrick, *Henry VIII*, (London 1968), p. 383.
14. *Henry Brinkelow's Complaint*, (1542), EETS, e series, 22, (1874), pp. 29–32.
15. William Turner, *A preservative, or triacle against the Poison of Pelagius*, (London 1551), BL 4256 a. 68. pp. 4–5.
16. *A Discourse of the Commonweal*, (1549), ed. M. Dewar, (Virginia 1969), p. 135.
17. J.E. Neale, *Elizabeth I and Her Parliaments*, (London 1953), I, p. 191.
18. Ibid.
19. *History of Parliament, The Commons 1558–1603*, ed. P. Hasler, (London 1981), I, p. 329.
20. R. Hooker, *Of the Law of Ecclesiastical Polity*, ed. C. Morris, (London 1909), vol. 1, bk. 3, pt. VIII, p. 309.
21. Ibid., vol. 2, bk. V, pt. 10, p. 37.
22. Ibid., vol. 1, bk. II, pt. 7, p. 268.
23. K. Wrightson, "The Puritan Reformation of Manners", PhD thesis, (Cambridge 1973), pp. 37–8.

4 "The Exceeding Lucre That They See Grow"[1]

Higher Profits and a Heightened Sense of Property

The disappearance of the architectural substance of the universal church, along with the patterns of communal and devotional mobility that revolved around it, constituted the most sudden and dramatic transformation in the character of the physical environment. At the same time, a more gradual but equally fundamental change was taking place in the shape of landholding and agrarian practice. By the end of the sixteenth century, there was a new balance of agricultural activity and tenure that visibly altered the face of the land. By that time, a significant area of the core arable farming regions, especially in the south and east, had been enclosed and brought under individual control, and the MP Henry Jackman could observe that the landscape was losing the character that had been typical of medieval times, "when the whole land lay in open fields, subject only to the hand of the husbandman".[2]

The redistribution of the church lands did not begin this process, but it had an important contribution to make, not just because of the enormous scale of property that became available following the Dissolution but also because of the manner in which it was marketed. It was demand led. Henry's appetite for acquiring church property was more than matched by that of his landholding subjects, and they were more capable of holding onto it. There was an understanding that the sweetener of the abbey lands would guarantee the success of Henry's break with Rome. In June 1534, the Spanish ambassador reported that the king was being advised to distribute the revenues of the church among the gentry, "that he may thereby gain the hearts and affections of his subjects".[3] It was a sound investment, and from the 1540s onward the gentry began to reap the dividends as a steady flow of ex–church property came onto the market.

The initiative in the transactions tended to come from the purchaser, who would have particular lands in view. The crown agents seem to have made little attempt to dictate the shape of the property to be bought. This would generally be determined by the customer, usually on the basis of location, where there was an opportunity to add to an existing farm or estate or to create a new one out of several adjacent manors. This "made for more compactness in the geographical pattern of landownership".[4] It thus contributed to the emergence of more coherent holdings, the development that was changing the balance of land tenure and reinforcing the idea of the integrity of property. So the redistribution of church lands was important not just as secularization but also as part of the process that was altering the shape of landholding. The holdings of medieval times had been comparatively dispersed in form and often united only in the person of their occupiers. Now lands were being consolidated, and self-contained and individually owned farms and estates were becoming the dominant feature of the countryside. A heightened sense of property was also encouraged by the nature of the conveyance. It was indeed "the largest transference of land since Domesday", but the transactions were different from the original conqueror's handout. Rather less than a quarter of the whole was simply granted by the crown in the old feudal manner.[5] The remainder, the great bulk of the church lands throughout England, were sold for cash, at a market price.

In his study of the West Riding of Yorkshire, R.B. Smith managed to put some interesting figures on the way that the redistribution of the church lands provided the final stage in the creation of a full-fledged land market. He noted that a land market already existed in embryo by the beginning of the sixteenth century, marked by the appearance of cash transactions, though "the land law had not been devised to allow for such transactions, and consequently the legal formulae available to record changes of this kind were clumsy adaptations of feudal forms".[6] Smith found that the volume of land exchanges began to increase significantly in the thirty years after 1510, with cash transactions multiplying at a much faster rate than feudal "enfeoffments for use". In the first five-year period, from 1510–1514, there were just nineteen transactions, of which eleven were for cash. The volume then began to rise sharply, mainly due to the increase in cash deals. By 1539, while the number of enfeoffments had roughly doubled, the number of cash deals had increased tenfold. It was at this point that the church lands came onto the market. And during the 1540s, with the powerful boost to supply, cash transactions soared, while enfeoffments declined to a few. In the five years between 1545 and 1549, out of two hundred and twenty-three transactions, three were enfeoffments and two hundred and twenty were for cash.[7]

As this makes clear, the redistribution of church land did not of itself create a land market, it consolidated a process that was already under way. The strength of the preexisting demand for land helps indeed to explain why the revolutionary act of the dissolution of the monasteries was carried through with so little significant problem. Many purchasers had very powerful practical motivations. There was a marked intensification of economic activity that created the basis of a land market before the sale of the abbey lands provided the final boost.

It is also evident what constituted the most important impetus for economic development. In the later years of the fifteenth century and the early years of the sixteenth, there was a notable change in the balance of agricultural activity and land use. Principally, it involved a substantial increase in the amount of land that was employed for sheep grazing, which was coming to be practiced on a much larger scale than in previous centuries. The incentive for this redirection of agricultural enterprise was the availability of a source of unprecedented levels of profit. This derived from the most conspicuous area of expansion in the early sixteenth-century economy. At the heart of it was the striking emergence of the cloth trade, to which the marked extension of sheep farming was the necessary accompaniment.

It was in various respects a new dimension in the economy. English and Welsh wool had long been renowned for its quality, and raw wool had been the principal exported commodity throughout the medieval period. After about 1440, however, exports of wool were falling. This was in part because the disruption of overseas markets had almost halved the price that could be achieved. More positively, there was the inducement of a strong home demand for cloth. Most groups of society contributed to this. After the population collapse of the fourteenth century, the relatively prosperous condition of the surviving labourers apparently left them more to spend on accessories like textiles. A contemporary noted that once upon a time the "servingman" was content with the coarsest kind of cloth, but "now he will look to have at least for the summer a coat of the finest cloth that may be got for money and his hose of the finest kersey".[8] In fact, the domestic consumption of cloth manufactures doubled in the second half of the fifteenth century.[9]

Wool now had an industrial as well as an agricultural application, and this was the basis for further commercial advance. Cloth manufacture was carried out mainly as a cottage industry in the countryside, and it could bring valuable by-employment to areas that were not best suited to productive farming. This was the case for instance in the West Riding of Yorkshire, where cloth making expanded rapidly among small farmers struggling to cope with poor soils and an over-wet climate. Thus an act of 1555 endorsed the necessity of cloth production around Halifax, which was "planted in the great wastes and moors where the fertility of the ground is not apt to bring forth any corn or good grass" and the people "altogether do live by cloth making".[10] In this respect it was a diversification that could tap underused resources. Daniel Defoe gave a good description of the beneficial process of regional specialization and exchange that this could generate. He found the West Riding in a state of prosperity that was entirely dependent on the cloth trade. "For, as to corn . . . they sow little . . . all of them employed in the clothing trade . . . their corn comes up in great quantities out of Lincoln, Nottingham and the East Riding, their black cattle and horses from the North Riding . . . their cheese out of Cheshire and Warwickshire".[11]

The market was able to expand to the full because the cloth product, like the raw wool, was found to be of sufficient quality to extend the trade overseas. English cloth producers were able to take advantage of the fact that the rise in population and economic activity, which characterized sixteenth-century Europe in general, had begun rather earlier on the continent. The demand in this instance came mostly from the urban middle classes and nobility, since the continental working classes appear to have been held below the level of consumers.[12] At any rate, by the end of the fifteenth century, cloth had become England's first significant manufactured export. And by 1513, the Venetian ambassador could hail English kerseys as "one of the most important foundations of trade in the world".[13]

A new awareness of economic opportunities was encouraged by the fact that these trends were both powerful and sustained. The expansion of the cloth trade was sufficiently marked and consistent to generate a new economic context. The productivity of cloth manufacture and the growth of cloth exports continued to advance strongly for at least three-quarters of a century, through to the climactic boom decade of the 1540s. The trade seemed to have everything in its favour. It was lightly taxed, and unlike the corn trade, it was not subject to export restraint when prices reached a certain level. Thus grazing became associated with an area of rising productivity. The existence of a steadily growing cloth market, which created manufacturing capacity as it went along, added to the inherent advantages that sheep farming held for landowners. Grazing was in general a less precarious exercise than arable husbandry, and it required only a minimum of labour at a time when labour costs were still at a relatively high level. A contemporary summed up these comprehensive advantages with the thought that the products of grazing involved "small charge and small labour" and had "free vent to be sold both on this side and beyond the sea at the highest penny".[14]

It was sheep farming, then, that promised the lowest possible overheads and the highest possible profits. This was an enticing combination. And it was certainly no accident that the first book of husbandry to be printed in English appeared in the middle of the period of expansion in the cloth trade and gave priority to a detailed analysis of the best methods of sheep farming because sheep were "the most profitablest cattle a man can have".[15] The phenomenon was widely recognized and was enough to tempt some of the larger landholders and manorial lords to resume direct management of the demesne. "They see that there is more advantage in grazing and breeding than in husbandry by a great deal . . . what should better encourage them thereto than to see them that do it become notable rich men by doing thereof in brief time".[16] There were indeed unprecedented opportunities for fast and consistent profits. It was a great inducement for manorial lords and farmers to enclose and take control of their lands because of "the exceeding lucre that they see grow".[17] The level of profit exceeded anything that had previously been available. So whereas the fifteenth century had been marked by a tendency for landlords to let out the demesne, the early sixteenth century saw a trend whereby some took it back in hand, and "gentlemen became graziers".[18] This was the basis of the first notable period of enclosure of the commons and conversion to pasture, which Robert Allen calls the landlords' agricultural revolution. It produced a significant alteration in the landscape of some of the most sensitive arable farming regions. "Between 1450 and 1525 about one tenth of the villages in the South Midlands were destroyed . . . as lords seized peasant land . . . the enclosures eliminated small-scale agriculture and represented an abrupt transition to capitalist relations".[19]

The enterprising landlord was not, however, the most important basis of the distinctive level of commercialization that English farming was to achieve during the sixteenth and seventeenth centuries. A greater force came from below. The seeds had been sown in the early fifteenth century, when the exceptional profits of sheep farming were not yet available and many landlords began to surrender the direct farming of their lands because it was difficult to sustain with advantage. The collapse of the population level in the fourteenth century had led to a decline in the demand for tenant land, an acute shortage of labour and a fall in the market price for arable produce. The first response of landlords was to exploit their customary rights over tenants more vigorously because the relative value of dues and services was now higher. But the accumulated resentment at this oppressive stance was a significant cause of the Peasants' Revolt of 1381, which in turn induced a change of policy. The steward of the Berkeleys recalled that it "was much occasioned by the insurrection of Wat Tyler, and generally by all the commons in the land" that the family began to lease out their estate.[20] "Then instead of manuring the demesnes with his own servants . . . this lord began to . . . tack in other men's cattle onto his pasture . . . and to sell his meadow grounds . . . let out by the year still more and more . . . sometimes at racked improved rents".[21] It became general in the fifteenth century for lords to lease out their demesnes.

For the first time, the position of the working farmer was significantly liberalized, putting the larger tenants in a good position to take advantage of rising market opportunities and extend their holdings. Jane Whittle concluded that "the significance of the period between 1440 and 1580 . . . lies in the freedom, prosperity and . . . the lack of landlordly interference experienced by the rural population, and in the fact that an economy generated by small landholders unburdened with heavy exactions by state or landlord could promote the development of capitalism".[22] In fact, a new class of more substantial, commercializing village farmers was created. The word "yeoman" came into vogue to describe them, while the word "husbandman" differentiated those who remained in the condition of the typical medieval smallholder, the "quarter virgater" with about thirty acres of land.

The socio-economic success of the yeoman farmers has sometimes been attributed, perhaps too readily, to the tendency of the Tudor courts to give greater security to traditional copyhold tenures.[23] In truth, many of the new arrangements were of a more modern, leasehold variety: "first adopted when a lord, lacking labour, farmed out the demesne to one or more tenants . . . who paid a market rent . . . such men were often graziers who possessed some capital, and belonged to the new age of commercial agriculture".[24] It was not the form of tenure but the *scale* of a farming enterprise that was the most crucial and general determining condition of an ability to succeed in the market. The typical rising yeoman farmer would have acquired leases giving him a holding of sixty acres or more. Hugh Latimer cited his father as an example. "A yeoman, and held no land of his own . . . tilled as much as kept half a dozen men. He had a walk for a hundred sheep, and my mother milked thirty kine".[25]

It was the depth and breadth of commercialization that made England different, dissolving the balance of the feudal economy more thoroughly than elsewhere. In the European perspective, a commercializing trend was also noted in the Netherlands and in some parts of Scandinavia, but "the extreme position is in England, where many landlords ceased to farm their land directly, but leased it to tenant farmers who cultivated it with the aid of wage labour".[26] In the high medieval period, the interdependence of the manorial estate and the village community had maintained an essential stability in agrarian economy and society. With the break-up of this combination, the base for commercial activity was markedly broadened. The labour market also became more liberalized, since the employer was now as likely to be a commercializing yeoman as a lord with a prescriptive right to services. At first, labourers were still able to command relatively high wages, which further increased consumer demand. It was a virtuous circle of expansion, creating the platform for the continued development of the cloth trade and sheep farming, which in turn further encouraged the consolidation and commercialization of holdings.

Thus, the advent of unprecedentedly profitable areas of farming gave an added level of incentive to individual initiative in the market, whether by the lord of the manor or by the more substantial yeoman farmer. The new power of the profit motive not only put land at a higher premium but also changed the way that the role of landholding was viewed. In the medieval context, land had been held as the basis of a stable social structure and used conditionally on behalf of the community. It now came to be seen more as personal property, to be used for profit and further acquisition. So in the course of the sixteenth century, the "conditional" concept of property that had prevailed during the Middle Ages began to change towards the "absolute" notion of ownership that we recognize today. Marc Bloch saw this indeed as the "one really striking transformation" in the agrarian history of early modern Europe, and he thought that it occurred in England but not elsewhere on the continent.[27] The general economic inducement for a more absolute sense of property was the persistence through the sixteenth and early seventeenth centuries of a wide and varied seller's market with an exceptionally strong rise in agricultural prices, as well as the existence of a broad class of commercializing farmers and traders able to use the market to its full extent for the accumulation of wealth and land. But there were also two more specific factors that actually determined and defined the development of an idea of absolute property, and these were indeed distinctive to the English kingdom. One was the assumption of a perceived right to trade and use property free of arbitrary restrictions and impositions. The other was the existence of a representative system peculiarly well suited to promoting this kind of freedom.

So the high level of commercialization and the distinctive rise of absolute property derived principally from the development of general economic and political freedoms. Too much emphasis has been put on the enhanced security of tenures. Robert Allen appears to place upon it the whole burden of explaining the "yeoman's agricultural revolution".[28] But economic trends were more significant. Although there was a tendency for the Tudor courts to give greater protection to customary tenures, only a small proportion of landholders came before them. The more crucial and general determining factor was to be found in the scope for market activity and the assumed freedom with which it was pursued.

The most extreme contradiction of socio-economic realities came from Alan Macfarlane. He sought to undermine the traditional assumption that the sixteenth century witnessed a transition from a peasant-based economy to a more individualist context. In trying to refute this, he ignored the principal elements of differentiation and periodization. He began from an artificially narrow definition of "peasantry" in order to argue that England had never possessed such a class. To convey the idea that medieval England was already an individualistic world, he cited a statute of 1290, which authorized the freeman to sell his land. But this had very limited application. Freemen were a tiny minority of landholders, and it remained extremely rare for land to be willed away. Such rights did not reflect the defining character of the medieval economy. Economic ethics pointed in the opposite direction, subordinating the individual to the corporate being, as seen in the communal system of farming that typified the time. In fact, it was not until 1530 that the general body of landholders were given the right to alienate property, in response to the commercialization of land that had been taking place since the late fifteenth century. Most of Macfarlane's evidence of individualistic activity comes from the late sixteenth and early seventeenth centuries, which he then conflates with the medieval period. It actually conflicts with his theory, for the end of the sixteenth century and the beginning of the seventeenth is precisely where we would expect to find a more commercialized society, following the distinctive economic transition that had taken place since the end of the fifteenth century, with the Tudor inflation and the enduring sellers' market, and the qualitative change in the scope of trading activity.

Macfarlane's ill-founded theory has generated some equally insubstantial responses. Since he suggested that the breakdown of the land–family bond was a measure of individualistic activity, other historians have sought to preserve a distinction with the medieval period by showing that such a bond still obtained at that time. The land–family bond debate has proceeded at great length without ever transcending the inadequacy of its terms of reference. It proved impossible to define a family for these purposes, and it is not clear why in any case it would indicate a disinclination to use the market.[29] The more coherent truth is that there *was* a transition to a more capitalistic economy in sixteenth-century England, when the loosening of manorial ties took effect and yeomen farmers were freed to take advantage of new levels of opportunity and profit, as the nature of marketing itself changed, acquiring a regional and international range, with specialization and intensified commercial practice.

The capacity to take advantage of the market conditions did not depend on any general reform of landholding arrangements, and there was no apparent demand for it. Feudal tenures merely mutated, sometimes imperceptibly, into more modern shapes, as economic circumstances dictated. The conditions of the sixteenth century indicated a need, or an opportunity to adjust the values of tenures, to keep pace with the increase in population and the even faster rise in agricultural prices. Landlords did not fail to take advantage of these trends. It appears that in general they managed to raise their income in line with inflation. "The rise in rents seems to have outstripped the rise in agricultural prices".[30] Feudal leases had tended to be for long terms and carried both rents and entry fines. The first response of landlords to the inflationary trend was to crudely raise fines. But by the end of Elizabeth's reign, there was a tendency to dispense with the entry fines and offer shorter leases at commercial rents. "Leasehold and copyhold were parting company, and the former was assuming its modern shape".[31]

Copyholders "for lives" were the most numerous of their class and the most characteristic of feudal tenures. They were also the least secure. They had evolved out of the basic, unfree "villein" tenure, which was subject to the will of the lord. Over the years, the villein established customary rights to a holding, which were copied into the manorial roll. But these copyholders never achieved individual control of their tenure. They had only the first right of refusal on renewal, and the entry fines were arbitrary. This often led to larger tenants taking over smaller farms. It should not be supposed that all or even most copyholders were so adequately protected that the will of the lord or the force of the market could not undermine them. "The landlord was free to put in a more substantial tenant, or to throw several properties together to make a larger leasehold farm".[32] In the agricultural sellers' market of the sixteenth century, a sixty-acre holding virtually guaranteed a profit, but the traditional medieval holding of a thirty-acre "yardland" did not produce a regular marketable surplus and was prone to failure. This was the context in which the most typical peasant holding began to disappear, absorbed in more commercial forms of tenure.

Copyholders "by inheritance" were better placed. Eric Kerridge established the view that by the later sixteenth-century, these tenures were upheld in the courts as if they were freeholds. Robert Allen confirms this, stressing the importance of the fact that the courts were prepared to enforce a cap on the entry fines for these copyholders.[33] Peter Bowden's survey of tenures also indicates that these copyholders now possessed effective security. The level of their rents and entry fines was fixed, so "by the end of the sixteenth century rising prices had given the copyholder an almost freehold interest in his land . . . helped to pave the way for outright purchase".[34] Such copyholders had little need to press for a change in the definition of property because the existing arrangements were working in their favour. Their holdings were secure and profitable, and, equally important, they often had neighbours whose holdings were less so and offered scope for takeover.

The freehold tenure itself did not carry quite the sense of absolute property that we recognize today. A freehold might be held in perpetuity and "by fealty only", free of arbitrary revision, but it still carried services and usually a rent. The advantage was that since the rent was small and fixed, it tended to become less meaningful with the passage of time and rising prices. So by the end of the sixteenth century, after the great Tudor Inflation, the limitations on freehold ceased to have any force, and it began to attain its modern sense.[35]

Thus, the force if not the form of tenures was changing, often in a way that provided greater security for some farmers. But the more crucial determining factors remained the conditions of the market and the capacity to exploit it, and the critical point was that not all peasant farmers could be successful. In fact, the ability of some to expand was often dependent on the failure of other, smaller husbandmen whose acreage was not sufficient for them to use the market to their advantage and who might therefore be obliged to sell their farms. They were then likely to lose their position as independent landholders and become labourers. Their land would be taken into the consolidation of medium-sized farms that was dictating the new shape of the agrarian context. The English rural world was beginning to acquire the characteristic structure that would come to distinguish it from its continental counterparts over the course of the next two centuries. It was a process of socio-economic polarization, by which the traditional and once ubiquitous medieval or peasant smallholder was gradually displaced and disappeared, on the one hand into the ranks of wage labourers and on the other, if fortunate, into the ranks of the more substantial, commercial yeoman farmers.

The emergence of "absolute" property was therefore less about technical changes in the form of property holding and more the result of economic developments and the assumption of commercial freedoms—a change in the ethics of how land and property should be used. This is best conceptualized by looking at the sharply differing attitudes that were displayed towards the process of commercialization. The new breed of agriculturalists and surveyors appearing in print tended to represent the gentleman farmer, and they offered straightforward, unashamed advice on how to maximize profits. The unproblematic acceptance of the profit motive is summed up in Fitzherbert's recommendation of sheep as "the most profitablest cattle a man can have" and in his advocacy of enclosure, on the same pragmatic basis, as simply the best way to make the most out of your sheep.[36] This was the first expression of a purely commercial perspective and was echoed by agricultural writers throughout the sixteenth century and into the next. Thomas Tusser was similarly convinced that pasture farming could bring a transforming level of profit. "Good husbands with grazing do purchase their lands". He embraced enclosure for similar reasons. "Then hedge them and ditch them, bestow thereon pence".[37] John Norden, at the beginning of the seventeenth century, also approved of enclosure as "the most beneficial course the tenants can take to improve their abilities".[38]

Full weight should be given to the force of the profit motive for these agriculturalists. It is sometimes obscured by the supposition that the main advance sought in the sixteenth century was an increase in rents. The potential for raising productivity was in any case the soundest pretext for higher rents, and in fact the consistent rationale of the farming books was to establish the best ways of securing good profits. The impetus was clear enough. The exceptional value of the cloth trade and sheep farming had led to the proposition that "sheep were the most profitablest cattle a man can have" and that grazing was a source of "exceeding lucre". This established the perception that, by targeted and selective practices, it was possible to raise income to new levels. Accordingly, Fitzherbert prescribed many of the approaches that would come to be regarded as the basis of improved farming method, such as water meadows, convertible husbandry, and the benefits of more manure with, of course, enclosure.

On the other side of the argument were those who for one reason or another were more concerned about the harmful consequences that the disruption of the pattern of landholding might have on the position of the smaller husbandman. In this camp was every governing regime, political and ecclesiastical, from the 1480s to the 1630s. All the great masters of Tudor government—Henry VII, Thomas Wolsey, Thomas Cromwell, William Cecil and Elizabeth I—seemed to have been convinced that on balance the process of enclosure for pasture was socially destructive. Leading churchmen from Hugh Latimer in the 1530s to William Laud in the 1630s took the same view.

The opposition to enclosure found its most celebrated moral advocacy at an early stage, in 1516, with Thomas More's image of sheep eating men.

> Your sheep that were wont to be so meek and tame and so small eaters, now as I hear say, have become so great devourers and so wild, that they eat up and swallow down the very men themselves. They consume, destroy and devour whole fields, houses and cities. For look in what part of the realm doth grow the finest and therefore dearest wool, these noblemen and gentlemen . . . not contenting themselves with the yearly revenues and profits that were wont to grow with their forefathers, not being content that they have rest and pleasure nothing profiting, yea, much annoying the public weal, leave no ground for tillage. They enclose all into pastures; they pull down houses . . . one covetous and insatiable

> cormorant . . . compass and enclose many acres of ground together within one pale or hedge, the husbandmen be thrown out of their own, or else either by course of fraud or violent oppression be put besides it, or by wrongs or injuries be so wearied that they be compelled to sell all.[39]

The strongest tendency among recent historians has been to play down the scale of early enclosures and to treat More's picture as melodramatic scare mongering. But Robert Allen's exhaustive study of the South Midlands shows that it was not far from the truth. "Early enclosures, especially those before the mid sixteenth century, frequently involved the destruction of villages and the expulsion of inhabitants as lords seized peasant land".[40] By 1525, about one-tenth of the villages in this sensitive area had been destroyed. This was taking place on a sufficient scale in the agricultural heartlands that it was no surprise that More and many others felt obliged to draw attention to the problem.

More's analysis is important not only in highlighting a socio-economic trend but also in offering a statement of the traditional moral dimension that was supposed to govern the agrarian economy. This was the "conditional" definition of property, which was to be displaced over the next century. More exemplified the medieval state of mind, reflecting a world where things were not expected to change much, and the true aim of economy was to sustain the existing balances, moral purposes and customary bounds of society. A basic rule was that all produce should be brought to an open market where a "just price" would be established. There were statutes that sought to prevent the diversion or monopolizing of sales for private gain. Thus any attempt by an individual to profit beyond a due level was subject to restraint, as detrimental to the common weal, for it was the good of the whole community, finding its unity in the charity of Christ, that was the guiding principle.

The essence of moral economy as spiritual wholeness is even better illustrated in the work of Robert Crowley. Like More, he was outspoken in his condemnation of the "greedy cormorants" who "enclosed from the poor their commons" and "would eat up men, women and children". These landlords were breaking the God-given balance of agrarian society. Crowley sought to remind them that their possessions were a stewardship from God, to be used for the common good, "not to hold from others the things that they have need of". Nothing so clearly underlines the moral imperatives of these traditional assumptions than Crowley's suggestion that they would have the right to do exactly what they liked with their property only in a world where God did not exist. "If there were no God then I would think it lawful for men to use their possessions as they listeth". Currently the farmers, graziers, rich butchers and gentlemen were acting "as if there were no God at all". By the same token, they were departing from their God-given place in the socio-economic order. Nothing so clearly illustrates the new force of the profit motive than Crowley's complaint that the profiteers were men with no proper, recognizable vocation because "they are doers in all things that any gain hangeth on".[41]

This moral perspective makes for a vivid contrast with the unproblematic endorsement of enclosure for commercial advantage that was put forward by the agriculturalists. The two shades of opinion were looking at different sides of the coin. Although both were recognizing that the profit motive was operating with a new level of intensity, the agriculturalists were embracing it on the basis of making maximum rational use of market conditions, while the moralists were focusing on the fact that almost by definition it would have a disruptive and detrimental effect on the interests of others.

Robert Crowley illuminates the essential features of the conditional view of property that was a characteristic part of the moral dictates of the medieval period. One crucial facet was that land and property was held as a trust or stewardship from God to be employed in line with the common good, and the individual was not at liberty to use it for excessive gain to the detriment of others. Importantly, Crowley also noted the other principal restraint that the conditional view of property placed on the rights of the individual. People were obliged to pay the king whatever taxes he demanded, even if they could ill afford it, for he was the sole judge of the overall public interest. "It is not for thee to define what great charges the king is at".[42] Crowley thus identified the twin pillars of the conditional view of property, and it was quite specifically in the overturning of these two precepts that the concept of absolute property was to be established.

NOTES

1. *A Discourse of the Commonweal*, (1549), ed. M. Dewar, (Virginia 1969), p. 118.
2. J.E. Neale, *Elizabeth I and Her Parliaments 1584–1601*, (London 1953), p. 342.
3. CSP Spanish, V, I, p. 256.
4. J. Youings, "The Disposal of Monastic Lands", in *The Agrarian History of England and Wales*, ed. J. Thirsk, (Cambridge 1967), IV, p. 340.
5. Ibid., p. 343; J. Youings, "The Terms of the Disposal of the Devon Monastic Lands", *English Historical Review, LXXX*, (1954), pp. 31–3; and S.M. Thorpe, "The Monastic Lands in Leicestershire on and After the Dissolution", Oxford B. Litt. 1961, p.133.
6. R.B. Smith, *Land and Politics in the Reign of Henry VIII*, (Oxford 1970), p. 218.
7. Ibid., pp. 218–9.
8. *A Discourse of the Commonweal*, ed. M. Dewar, pp. 81–2.
9. J.H. Lander, *Government and Community: England 1450–1509*, (London 1980), p. 15.
10. *Statutes of the Realm*, IV, 2 & 3, Philip and Mary, cap. 13.
11. Daniel Defoe, *A Tour Through the Whole Island of Great Britain*, ed. P. Rogers, (Harmondsworth 1971), p. 496.
12. W. Minchinson, "Patterns and Structures of Demand", in *The Sixteenth and Seventeenth Centuries*, ed. C. Cipolla, (Hassocks 1977), pp. 99, 131.

13. Venetian State Archives, cited in F. Braudel, *The Mediterranean*, (London 1972), 1, p. 213.
14. *A Discourse of the Commonweal*, ed. M. Dewar, p. 119.
15. Fitzherbert, *Book of Husbandry*, (1523), ed. W. Skeat, (London 1882), BL Ac. 9935/8, p. 42.
16. *A Discourse of the Commonweal*, ed. M. Dewar, pp. 53–4.
17. Ibid., p. 118.
18. Ibid., p. 20.
19. R.C. Allen, *Enclosure and the Yeoman*, (Oxford 1992), pp. 13–4.
20. John Smyth of Nibley, *The Lives of the Berkeleys*, (1618), ed. J. Maclean, 3 vols., (Gloucester 1883–6), II, p. 5.
21. Ibid., pp. 5–6.
22. J. Whittle, *The Development of Agrarian Capitalism*, (Oxford 2000), p. 315.
23. R.C. Allen, *Enclosure and the Yeoman*, p. 14; E. Kerridge, *Agrarian Problems in the Sixteenth Century and After*, (London 1969).
24. H. Brailsford, *The Levellers and the English Revolution*, (Nottingham 1983), p. 438.
25. Hugh Latimer, "First Sermon Preached Before King Edward", 1 March 1549, *The Works of Hugh Latimer*, ed. G.E. Corrie, The Parker Society (Cambridge 1844–1845), I, p. 101.
26. W. Minchinson, "Patterns and Structures of Demand", in *The Sixteenth and Seventeenth Centuries*, ed. C. Cipolla, p. 98.
27. M. Bloch, *Land and Work in Medieval Europe*, trans. J.E. Andrew, (London 1967), p. 49.
28. R. Allen, *Enclosure and the Yeoman*, pp. 66–9.
29. For the debate, see Z. Razi, "Family, Land and the Village Community in Late Medieval England", *Past and Present 93*; G. Sweenavasun, "The Land–family Bond in Earls Colne 1550–1650", *Past and Present 131*; and R. Hoyle, "The Land–family Bond in England", *Past and Present 146*. Macfarlane's theory is in *The Origins of English Individualism*, (Oxford 1979).
30. M. Overton, *The Agricultural Revolution in England*, (Cambridge 1996), p. 170.
31. P. Bowden, "Agricultural Prices, Farm Profits and Rents", in *The Agrarian History of England and Wales*, ed. J. Thirsk, IV, p. 686.
32. Ibid., p. 685.
33. R. Allen, *Enclosure and the Yeoman*, p. 69.
34. P. Bowden, "Agricultural Prices", in *The Agrarian History of England and Wales*, IV, p. 684.
35. Ibid.
36. Fitzherbert, *Book of Husbandry*, pp. 42, 77.
37. Thomas Tusser, *A hundred good points of husbandry*, (London 1557), verses 17, 68.
38. John Norden, *The Surveyor's Dialogue*, (1618), ed. M. Netzoff, (Farnham 2010), p. 90.
39. Thomas More, *Utopia*, (1516), ed. J. Warrington, (London 1978), p. 26.
40. R.C. Allen, *Enclosure and the Yeoman*, p. 14.
41. Robert Crowley, "The Way to Wealth" and "An Information", *The Select Works of Robert Crowley*, EETS, e series, XV, (London 1872), pp. 132, 142–4, 157.
42. R. Crowley, "Epigrams", *Works of Robert Crowley*, EETS, e series XV, (London 1872), p. 67.

5 Enclosure and Consolidated Holdings

The Break-up of the Communal System

The free maximization of profits was the basic motive force of individualist farming and the most powerful dissolvent of the moral and communal ethic that had guided economic affairs in medieval times. The characteristic structural forms encouraged by commercialization were the enclosure and engrossing of holdings. This involved another notable change in the physical environment, as well as a reordering of the agrarian scene. It occurred first in the central southern region, where it had highest visibility. Already by 1558 a traveller could note that the balance of the landscape had altered and no longer presented the open face that had characterized the medieval countryside. "The fields are all enclosed with hedges . . . and . . . trees, to such an extent that in travelling you think you are in a continual wood".[1]

The disruption to the physical shape and communal habits of the countryside is illustrated by the problems that arose with the traditional Rogation-tide perambulations. To describe this takes us straight back into the world of open fields and processions, if only to visualize how it was being broken up. Keith Thomas noted that the "corporate manifestations of the village community" fell out of use less from rationalist objections to externalism and more "because of the social changes that broke up the old community, and physically impeded anything so cumbersome as a perambulation". The old ritual was well suited to "open field country, but enclosures and cultivation led to the destruction of the old landmarks and blocked the rights of way".[2] This encapsulates the force of the change and is borne out by other evidence. Ronald Hutton confirmed this as the principal difficulty affecting the perambulations in the reign of Elizabeth. Complaints against the perambulation as idolatrous were rare. The broader problem, with hundreds of examples, was more practical: "because of squabbles between the minister and the parish elite, because the route had been blocked by a new wall, hedge or ditch, or because of the lack of interest". These conflicts continued into the reign of James I, when "those living near the route sometimes blocked it with enclosures, or refused to provide customary hospitality to the walkers".[3] The tension reflected a fundamental rebalancing of socio-economic ethics. It was echoed in other developments at the same time, like the termination of the traditional liberty assumed by May Day revellers to raid estates and farms for timbers. Where once the community had exercised residual rights in the land, there was now a stricter assertion of absolute property.

If these transitions have received less than their due recognition in recent historiography, it is partly because of the controversy over the effects of enclosure on the position of the open field smallholder. The traditional supposition was that the first important phase of enclosure took place from the late fifteenth to the early seventeenth centuries and that this accounted for the contemporary outcry against it for driving the arable husbandman off the land. The most famous interpretation of this was by Richard Tawney, who characterized the process as the expropriation of peasant farmers by powerful landlords.[4] More recent opinion has been reluctant to accept the picture of ruthlessness, and Mildred Campbell concluded that "we no longer regard it as an historical calamity visited on the poor by the rich".[5] To further defuse the idea, some historians have suggested that only a small proportion of enclosure took place in that early period and that the first main phase was focused on the seventeenth century alone.[6]

There are many reasons to distrust these attempts to diminish the effects of enclosure in the late fifteenth and sixteenth centuries. Opposition to Tawney often stemmed from a dislike of his "picture of early capitalism as cruel and greedy".[7] But to reject Tawney's view is also to dismiss the persistent observations of sixteenth-century contemporaries that enclosure was indeed depopulating. Some historians went so far as to suggest that the outcry was "hysterical exaggeration", though it clearly represented the strongly held opinions of people who were actually there. In fact, the minimization of the incidence of sixteenth-century enclosure can be achieved only by exploiting the inadequacy of the statistical data. The sources do not admit of a precise computation of enclosure in the period. The practice generated widespread controversy, but no systematic record. The findings of Wolsey's commission of 1517 have been used as ammunition, but they are partial in every respect. "The surviving returns . . . are so incomplete that no conclusions about the extent of enclosing or engrossing can be built upon them".[8] Even such results as were obtained were far from reliable. Returns could be skewed by the corruption or intimidation of juries. The commissioners complained that the juries were often packed with the servants of the culprits. Jurors might in any case be disinclined to cross the local lord. One jury in Leicestershire declared that William Ashby had enclosed one hundred twenty acres, whereas in fact it is known that he had enclosed a thousand acres and depopulated the entire manor.[9] And in terms of the crucial issue of the amount of arable being lost, the intended land use was not always transparent. Land promised for tillage did not necessarily materialize. There was a recognized trend by which land was enclosed ostensibly for tillage, then quickly turned to pasture.[10] Another familiar ploy was to enclose many acres, then "make one furrow and sow that, but the rest they till not, but pasture with their sheep".[11] There are also examples of enclosure that are known to have taken place but eluded the commissions or any other record. The enclosure of the fields of Tetchwick, Buckinghamshire, by 1519 escaped official notice. And Peter Brandon talks of the "silent revolution" by which most of the open fields of Sussex were enclosed during the sixteenth and early seventeenth centuries, "un-chronicled in surviving documents. Fortunately the landscape itself has much to tell".[12] Recent work suggests that, "we may have understated the degree of rural waste enclosure in the sixteenth century" and that "enclosure could take place on a huge scale, without leaving much trace in the public records".[13]

The clear contemporary evidence that we possess indicates that the traditional assumption was correct and that the decisive period for the first main phase of enclosure was between the late fifteenth and early seventeenth centuries. Given the exceptional profitability of sheep farming from the mid-fifteenth century, there was every incentive for the more substantial farmer to create a compact and controlled area where he could manage his stock and his land with maximum efficiency. The benefits could be considerable. For the early Tudor period, it has been estimated that when farming land was enclosed, its annual value increased by 31%, and the annual value of land enclosed for pasture exceeded that of land enclosed for arable by 27%.[14] This assessment was affirmed by the best known of surveyors, John Norden, at the beginning of the seventeenth century. He thought that enclosure was "the most beneficial course the tenants can take to increase their abilities, for one acre enclosed is worth one and a half in common".[15] By 1641, Henry Best believed that enclosing his pasture had increased its value threefold.[16]

It would indeed have been surprising if the first main spate of enclosure had not come when the incentives were greatest and when sheep farming gave landholders the opportunity "to become notable rich men by doing thereof in brief time".[17] Robert Allen's study of the South Midlands reinstates the importance of the first marked phase of enclosure from the late fifteenth century to the mid sixteenth century. "Between 1450 and 1525 about one tenth of the villages in the South Midlands were destroyed . . . as lords seized peasant land".[18] Furthermore, the process had precisely the effects that the Tudor critics claimed—the early phase in particular did lead to depopulation. "Enclosure before the mid-sixteenth century had a dramatic—negative—impact on population. Circa 1550, the average population density of villages already enclosed was only half the density of those still open . . . the Tudor conviction that enclosure caused depopulation was fact, not fancy".[19] As recent opinion confirms, "there is growing evidence of enclosure before mid-century".[20]

These studies simply underline the observations of the most authoritative of contemporary sources, *A Discourse of the Commonweal*, a judicious assessment of the state of the English economy at the midpoint of the sixteenth century. It presents a rounded view of the argument about enclosure. Both the husbandman and the gentleman have their say, and everyone agrees that a disproportionate amount of enclosure has taken place in recent decades—the only thing in debate is the effect. The arbitrating voice of the Doctor strikes an interesting balance between accommodating the force of the profit motive and recognizing its more harmful results. He absolves enclosure from being the main cause of high prices but notes that it was often done by force, intimidation or deceit, and it was certainly to be blamed for undermining the position of the independent husbandman. Furthermore, it was clearly regarded as having reached a critical point in the years leading up to 1549: "[I]f that kind of enclosure do as much increase in twenty years to come as it has done twenty years past it comes to the great desolation and weakening of the strength of this realm".[21]

So the crescendo of public protest in the late 1540s against enclosure and its associated practices was by no means the exaggerated emotional reaction that it has sometimes been painted. The Protectorate government, under the Duke of Somerset, received anti-enclosure petitions from all over the country, and the official response was the same as that of other Tudor administrations—to issue commissions of enquiry and attempt to arrest the process of enclosure because it was believed to cause high prices and depopulation. If Somerset's policy has received shorter shrift than that of other Tudor regimes, it is because the agenda of many recent historians has been to overturn the traditional idea of The Good Duke as a visionary idealist and champion of the poor. Earlier historians, like W.K. Jordan, regarded him as ahead of his time: "a remarkable and magnanimous man".[22] But the modern generation, having misled themselves into the improbable theory that the past cannot prefigure the future, could not allow Somerset's advanced reputation to stand. So by the 1970s and 1980s, The Good Duke was being presented as a fraud and an incompetent.[23] The net result of this revisionist interpretation was to suggest that Somerset invented the problem of enclosure to curry favour with the poor by pretending to offer a solution.[24]

A more logical analysis is available: There was a real grievance, and Somerset was making an earnest attempt to address it. Even if historians had been correct in supposing that enclosure for pasture was not at its peak in the late 1540s, the process was cumulative, and the hardship became more marked in times of economic depression. We should also recall that "enclosure" served as an umbrella term for a variety of associated practices. The essence of the matter was the consolidation of individual properties. There was an ongoing trend of "engrossing"—the absorption of smallholdings into larger farms. Another aspect that was prevalent at the time was the less formal procedure of "encroachment". Jennifer Loach noted "the encroachment on commons and wastes and the extinction of tenants rights over common land—as entrepreneurs jostled for limited resources".[25] Those who lost out were the cottagers, with no more than customary rights on the commons and no capacity to resist the determined ambitions of the larger landholders. Somerset actually seems to have identified a tangible injustice with some precision. In July 1549, he was reported to have told the Privy Council that "the peasants demands were fair and just, for the poor people who had no land to graze their cattle ought to retain the commons, and the noble and rich ought not to seize and add them to their possessions".[26] Whether or not the general economic effects of enclosure were harmful or beneficial is another question, but Somerset was certainly addressing a live social problem. Evidence has emerged recently confirming that he showed a genuine sense of justice in dealing with the protests of the commons.[27]

Ironically, his economic policy was, in one sense, backward-looking. The area where Somerset took an obviously advanced position in the medieval-to-modern balance was his tolerationist stance. His campaign against enclosure, however, derived mainly from the tradition of moral economy and stood against the incoming tide of capitalist marketing. The enclosure commission and the jurors were briefed as if it were a crusade to halt the advance of evil. They were enjoined to a "Godly act", remembering that the tenets of Christianity "forbiddeth the rich to oppress the poor". An interesting feature of the instructions is the sense they give of the power of the enclosers. These were men who could dominate their neighbourhoods and evade the law. The jurors were encouraged to stand firm. "For God, the King and the commonwealth, if ye serve them truly and faithfully, as they may be able to defend you against the Devil, the world and private profit".[28]

Somerset's approach was distinguished by the level of conviction and the readiness to act. When the commissions seemed powerless to overcome landlord influence and packed juries, he kept his word and proceeded to coerce the offenders. Part of the forfeited estates of the Duke of Norfolk and Thomas Seymour were ploughed up, as was the park of the Earl of Warwick. In view of the recent tendency to treat Somerset as incompetent, it is strange that in the long history of Tudor enclosure regulation, this was probably the only time when the gentry felt their interest and freedom in land encroachment to be seriously challenged.

The vigorous reaction of both government and people against enclosure and engrossing in the late 1540s was thus heightened by the sense that, at a time of hardship, the old assumptions of moral restraint and equitable provision were being pushed aside. Joan Thirsk has noted that "[t]he complaints were all of one kind, recording the resentment of the many at the selfish ambitions of the few. And they echoed from Solway Firth to the East Anglian Fens, from the Essex Marshes to Land's End". The selfish profiteering of the few could be seen as the cause of the problems, and the reassertion of the common good could be presented as the answer. But there was really no way back. A basic change was taking place in the balance of agrarian economy and society. The tensions "mirrored two different ways of farming—the individualist and the communal systems".[29] The old world where the open fields and commons were worked communally and cooperatively was giving way to a new world where consolidated farms were worked individually and commercially.

The controversies of the late 1540s reflected a watershed in this respect. It was the moment when the force of change became sufficiently evident to provoke widespread public anger. This was the time when people began to feel that the earth was moving beneath the small husbandman, and towards the substantial commercial farmer. As Robert Allen concluded, in that first phase "enclosure played a more important role in destroying the English peasantry than historians have been prepared to allow".[30]

There was a hiatus in the process of enclosure during the middle of the sixteenth century. This was not surprising, since the long expansion of the cloth trade had come to a pause, and the later 1550s saw the only five-year period in the Tudor century when the steady rise in population received a setback. By the 1570s, the agricultural economy was recovering strongly, and enclosure resumed. In fact, in the South Midlands, over the period 1450–1750, the really peak years of enclosure were between 1575 and 1650, when 60% of it took place.[31] It tailed away sharply in the second half of the seventeenth century, when the trends of population and prices were generally static.

The effect on the Midlands, the heartland of English tillage, was of special significance. Across the great Midland Plain, where it mattered most, there was undoubtedly serious depopulation and unemployment.[32] Here, where the evidence is clearest, it appears that 36% of enclosure took place between 1485 and 1550, and 40% in the early decades of the seventeenth century, with only 17% in the second half.[33] W. G. Hoskins suggested that a more relevant and instructive measure than the acreage enclosed is the number of places that were affected. And in his study of Leicestershire, he found that between 1485 and 1607, "more than one in three communities suffered complete or partial enclosure or conversion to grass". He concluded that "this was a shattering loss for the community concerned, at a time when pressure on land was rising".[34]

In another Midland county, Bedfordshire, "through the medieval period the size of the holdings, and their scattered disposition among the open fields remained remarkably static. But by the middle of the sixteenth century the situation was transformed . . . over some fifty years there is abundant evidence that piecemeal enclosure was occurring in the open fields".[35] And in the heel of English arable country, the picture was much the same. We have noted Peter Brandon's identification of the "silent revolution" by which the great southern flank of open field country in Sussex was enclosed in the late sixteenth century. Even when there was a genuine process of exchange and compensation, it still resulted in a change in the balance of agrarian practice. It was "the super-session of the old communal system".[36] At the other end of the kingdom, Bill Shannon has found that the great majority of the lowland wastes of Lancashire were enclosed or partitioned in the sixteenth century and the early decades of the seventeenth. The process came to an end "just before the time of the Civil War".[37]

Thus, if we avoid an arbitrary, simplistic comparison between separate centuries and focus on the more substantial chronological phases between the last decades of the fifteenth century and the early decades of the seventeenth, it becomes possible to suggest what the process of enclosure meant in terms of the most notable general circumstances, like major political changes and distinctive, long-term economic trends. The steady rise in population and price levels that continued through to the 1640s created the seller's market that tended in every way to encourage the enclosing and consolidation of holdings, which was at once the main vehicle and the principal consequence of the development of commercial farming.

The engrossing landlords and farmers could allow the market to work for them. It was a common sequence for a small farmer to fall into arrears with rents and thus be unable to avoid selling. The raising of an entry fine, even if not deliberately exorbitant, could be the last straw for a smallholder who had already fallen foul of bad harvests or market condition and who had become a prime candidate for being bought out. In fact, it is clear what scale of farming enterprise was needed to avoid vulnerability and give reasonable assurance of success in almost any circumstances. The best placed were those whom Peter Bowden has usefully described as the "chief farmers", with sixty acres or more on an essentially arable holding. The advantages they held are not hard to identify, given the consistent seller's market that prevailed in the sixteenth century. Bowden states the position graphically and dispassionately. "It was somewhere between 60 and 100 acres when general crop failure meant an increase in net income to the producer" because he had "a proportionately greater market surplus".[38] He could wait until the price rose, and the capacity to sell his surplus at an inflated level could more than compensate him for a bad harvest. The small farmer on the other hand was disadvantaged in every respect. In a good harvest year, he might have a small surplus to sell but had to be satisfied with selling it cheap. In a bad harvest year, he would probably have a shortfall and be forced to make it up by buying-in dear. Bowden notes the unfortunate consequence: "[S]everal bad harvests could hardly fail to bring ruin to these small cultivators".[39]

Margaret Spufford's study of Cambridgeshire confirms that the farms that were not reduced in number or size between 1560 and 1636 were the larger copyholds of sixty acres or more. Farmers with fewer than forty-five acres, aiming essentially at self-sufficiency, were most likely to disappear.[40] This illuminates a situation where traditional smallholdings were vulnerable to a process of decline in the prevailing market conditions. "The 'typical' medieval holding was no longer a viable unit in the price rise of the sixteenth century. In two parishes the engrossing movement and the disappearance of the small landowner have been shown to be phenomena which were accelerated in the late sixteenth and early seventeenth centuries. In one of them it was possible to show with precision that the last half-yardland unit disappeared in exactly those years of dearth when prices were the highest so far recorded".[41] Most strikingly, Spufford identifies "the arch engrosser", the farmer who had managed to incorporate the lands of fifteen other men, largely through forfeited copyholds.[42] Once this process had begun, it carried an inexorable momentum. So the old world where the small husbandman occupied the land in concert with his neighbours was coming to an end, and the world of the individual landowner, employing labourers on a commercial basis, was coming into being.

Care should always be taken in applying the word "peasantry", and we probably need to question Robert Allen's characterization of the success of the chief farmers as the "consolidation of the peasantry".[43] It was more the emergence of a new class—the substantial yeoman farmer, generating a significant surplus, on a minimum sixty acres. This was often achieved at the expense of the class that we would normally think of as the peasantry—the traditional medieval husbandman, with a typical holding of perhaps thirty acres and therefore under threat in the prevailing market conditions. The yeoman was of a specifically different status from the husbandman.

The policies and principles that had sought to safeguard the interests of the smallholder were also losing their force. The mid-seventeenth century was a watershed in political and economic ethics. Certain crucial aspects of the old assumptions of moral economy failed to survive the Interregnum. Over the preceding one hundred fifty years, the public authorities had consistently condemned enclosure and tried to restrain it, in the belief that it caused depopulation and the destruction of tillage. But the parliamentary challenge to the crown in the 1640s was the triumph of those classes—the gentry, yeomen farmers and merchants—that had long fretted against the restrictions on their freedom in the market. They were not inclined to hinder the process of enclosure now that they could influence events directly.

So the parliamentarian movement did not embrace the commoners' cause against enclosure. Even the Levellers were essentially in favour of market freedoms, and they adopted an anti-enclosure stance only briefly and belatedly, in response to their contacts with the common soldiers in the New Model Army. The Diggers were the one group consistently opposed to enclosure, but with no practical effect. The people in power, the MPs, were overwhelmingly on the side of the enclosers. The last bill attempting to restrain the process was introduced into the Commons in 1656 and thrown out without a second reading. In rejecting the bill, MPs proclaimed the enthronement of the new market orthodoxy. Master of the Rolls William Lenthall said, "he never liked any bill that touched upon property". Edmund Fowell called it "the most mischievous bill that ever was offered to the House".[44] This has rightly been called a turning point in the history of enclosure.[45] For more than one hundred fifty years, the aim of legislation had been to restrict the process, but henceforth parliament would legislate only to encourage it.

Even the moral injunctions of the clergy were falling silent. The last clerical tirade condemning the practice as irreligious was in the pamphlets of John Moore between 1653 and 1656.[46] In reply, another cleric, Joseph Lee, asserted that enclosure was "both lawful and laudable" and gave an explanation of why it was becoming normative. He said that it was natural for men's actions to be determined by the calculation of their own profit or advantage and that he knew of no law, divine or otherwise, that would deny them that right. He asserted furthermore that it was only the increased yield from enclosed fields that made corn currently so cheap.[47] Enclosure required no future defence. In the Interregnum decades, it was embraced in the concept of improvement formalized in the writings of agriculturalists like Walter Blith and Sir Richard Weston. "Any hindrance to enclosure was allocated to the world of backwardness".[48] The criticism died away. "After mid-century . . . the affirmation of a moral order where economic activities were means to social ends—God's and man's—fell from public view".[49]

It is therefore crucial to get the periodization clear. It clarifies the relationship between the enclosure movement and the change in economic ethics. The mid-seventeenth century was a watershed not merely because of the substantial amount of enclosure that had taken place by that time but also because it had occurred in a context and a climate that gave it a particularly disruptive force. In the one hundred fifty years before 1640, enclosure was essentially a law unto itself. It took encouragement from the conditions of the market, and the developing assumptions of individualist property, but it operated outside the regulatory provisions of the day and against the better judgement of the political and ecclesiastical authorities. Not all enclosure was depopulating, but it always undermined the communal structures that had been the prevailing pattern of medieval agriculture, and it contradicted the moral ethos that had been supposed to govern economic affairs. Thus the pre-1650 period has a distinctive significance in the history of enclosure because it was then most challenging to traditional practices and relations. After 1650, however, the status of enclosure was in many ways reversed. The legislative process no longer sought to restrain it but to facilitate it. It was in line with the newly triumphant rights of property, and it now reflected the dominant shape of agriculture on the ground. In effect, enclosure had become a different kind of process. It was normative rather than subversive.

Since the mid–seventeenth-century revolution changed the climate in such critical ways, we should be wary of supposing a significant continuation of the moral, or peasant, economy thereafter. E.P. Thompson has written of the "moral economy of the poor" in the eighteenth century and has related it to the earlier ethic, on the basis that the rioting crowds were trying to enforce the supply of grain in the same way as the government Books of Order of the 1580s and 1630s.[50] But the balance of forces had changed. These protests reflected the indignation of a working class that had already lost its independent position in the fields, not the true moral economy of communal rights, as a God-given principle of economic solidarity and restraint that had been embraced by medieval society as a whole and acted upon by the political and ecclesiastical authorities until the 1640s. The latter concept had been undermined beyond recall by the middle of the seventeenth century.

The same reservation must be made against reading too much into the survival of some of the outline structures of open field agriculture. Although many of the open fields and common lands did not finally disappear until the early nineteenth century, it should not be assumed that they retained the same undifferentiated character over the entire period 1500–1800. Robert Allen suggested that by 1800 the open fields were registering over 80% of the improvement in yields attained by those that were enclosed. But it is likely that by this time open field culture had departed some way from its original ethos and was working on a set of assumptions that were significantly more commercial and less communal than they had been in 1500. Most importantly, even where the fields remained physically unenclosed, the process of consolidation of larger farms and flocks in individual hands could still continue, and this was the essence of the change.

So the crucial transition had occurred by the second half of the seventeenth century, when communalism ceased to reflect the orthodox economic philosophy and the dominant shape of farming on the ground. Common field agriculture no longer defined the character of the agrarian scene as it had done in medieval times. By the end of the seventeenth century, it was becoming clear that England was witnessing the disappearance of the small, independent husbandman, who was being edged out of the picture, displaced in a distinctive polarization between substantial commercial farmers on the one hand and agricultural labourers on the other. Keith Wrightson has written the best known summary of the process, but the fact of it is generally acknowledged. The transformation was not complete until the beginning of the nineteenth century, but the decisive change had already occurred before the end of the seventeenth. It can be measured in terms of the rising proportion of landless people. The best estimates are that in the mid-sixteenth century 12% had no land, by the later sixteenth century it was 20–25% and by the 1680s it was almost 70%.[51] There had been a fundamental rebalancing of agrarian society. At the beginning of the sixteenth century, the independent smallholder had been in a large majority in the countryside, but by the end of the seventeenth century, there was an equally large majority of landless labourers. The world of the medieval husbandman was drawing to a close, and England had acquired a rural proletariat. Maurice Keen has come closest to capturing the effect: "The squire, the yeoman and Hodge the farm labourer are clean different figures from the lord, the reeve and the quarter virgater".[52]

NOTES

1. Estienne Perlin, *Description of the Realms of England and Scotland*, (1558), ed. R Gough, (London 1775), p. 25.
2. K. Thomas, *Religion and the Decline of Magic*, (Harmondsworth 1971), p. 74.
3. R. Hutton, *The Rise and Fall of Merry England*, (Oxford 1994), pp. 142–3, 175.

4. R.H. Tawney, *The Agrarian Problem in the Sixteenth Century*, (London 1912).
5. M. Campbell, *The English Yeoman*, (London 1960), p. 86.
6. M. Overton, *The Agricultural Revolution in England*, (Cambridge 1996), p. 148.
7. E. Kerridge, *Agrarian Problems in the Sixteenth Century and After*, (London 1969), p. 15.
8. J. Thirsk, in *Agrarian History of England and Wales*, ed. J. Thirsk, (Cambridge 1967), IV, p. 241.
9. W.G. Hoskins, *The Age of Plunder*, (London 1976), p. 69.
10. *A Discourse of the Commonweal*, ed. M. Dewar, (Virginia 1969), p. 51.
11. "Instructions to the Enclosure Commission, June 1548", in *Tudor Economic Documents*, ed. R.H. Tawney and E. Power, (London 1624), I, p. 41.
12. M. Reed, *The Buckinghamshire Landscape*, (London 1979), p. 152: P. Brandon, *The Sussex Landscape*, (London 1974), p. 148.
13. R. Hoyle, *Custom, Improvement and Landscape*, (Farnham 2011), pp. 14–5.
14. J.D. Mackie, *The Earlier Tudors*, (Oxford 1966), p. 450.
15. J. Norden, *The Surveyor's Dialogue*, (1618), ed. M. Netzloff, (Farnham 2010), p. 90.
16. Henry Best, Surtees Society XXXIII, *Rural Economy in Yorkshire*, (1641), p. 129.
17. *A Discourse of the Commonweal*, (1549), ed. M. Dewar, pp. 53–4.
18. R.C. Allen, *Enclosure and the Yeoman*, (Oxford 1992), pp. 13–4.
19. Ibid., pp. 42, 48.
20. R. Hoyle, *Custom, Improvement and Landscape*, p. 12.
21. *A Discourse of the Commonweal*, ed. M. Dewar, p. 49.
22. W.K. Jordan, *Edward VI*, (London 1968), vol. I, pp. 172–5.; A.F. Pollard, *England Under Protector Somerset*, (London 1900).
23. M.L. Bush, *The Government of Protector Somerset*, (London 1975); D.E. Hoak, *The King's Council in the Reign of Edward VI*, (Cambridge 1976); J. Guy, *Tudor England*, (Oxford 1988).
24. J. Guy, *Tudor England*, pp. 201, 207–8.
25. J. Loach, *Edward VI*, (Yale 2002), p. 60; R. Manning, *Village Riots*, (Oxford 1988), p. 36.
26. CSP Spanish, IX, p. 395.
27. Somerset displayed a willingness to negotiate with protesters in good faith, unlike other sixteenth-century rulers: D. MacCulloch, *Tudor Church Militant*, (London 1999), p. 47.
28. *Tudor Economic Documents*, ed. R. Tawney and E. Power, I, pp. 39–44.
29. J. Thirsk, *The Agrarian History of England and Wales*, IV, pp. 205–6.
30. R.C. Allen, *Enclosure and the Yeoman*, p. 54.
31. Ibid., pp. 28–9.
32. J. Thirsk, *The Agrarian History of England and Wales*, IV, p. 240.
33. M. Overton, *The Agricultural Revolution*, p. 148.
34. W.G. Hoskins, *The Age of Plunder*, (London 1976), p. 70.
35. P. Bigmore, *The Bedfordshire and Huntingdonshire Landscape*, (London 1979), pp. 154–5.
36. P. Brandon, *The Sussex Landscape*, pp. 146–8; and Brandon, "The Common Lands and Wastes of Sussex", PhD thesis, (London 1963), pp. 307–10.
37. Bill Shannon, "Approvement and Improvement in the Lowland Wastes of Early Modern Lancashire", in *Custom, Improvement and the Landscape*, ed. R. Hoyle, pp. 191, 201.
38. P. Bowden, "Agricultural Prices, Farm Profits, Rents", *The Agrarian History of England and Wales*, ed. J. Thirsk, (Cambridge 1967), IV, p. 659.

39. Ibid.
40. M. Spufford, *Contrasting Communities*, (Stroud 2000), pp. 90–2, 118–9.
41. Ibid., p. 165.
42. Ibid., pp. 80–1.
43. R.C. Allen, *Enclosure and the Yeoman*, p. 66.
44. *The Parliamentary Diary of Thomas Burton*, ed. J.T. Rutt, (Colburn 1828), I, pp. 175–6.
45. H. Brailsford, *The Levellers and the English Revolution*, (Nottingham 1983), p. 426.
46. John Moore, *The Crying Sin of England of Not Caring for the Poor*, (London 1653); *A Scripture Word Against Enclosure*, (London 1656).
47. Joseph Lee, *A Vindication of a Regulated Enclosure*, (London 1656), pp. 7–9.
48. P. Warde, "The Idea of Improvement", in *Custom, Improvement and the Landscape*, ed. R. Hoyle, p. 142.
49. J. Appleby, *Economic Thought and Ideology in 17th Century England*, (Princeton 1978), p. 70.
50. E.P. Thompson, "The Moral Economy of the English Crowd in the 18th Century", *Past and Present 50*, (1971), pp. 79, 108.
51. M. Overton, *The Agricultural Revolution*, p. 179. For Keith Wrightson's overview, *English Society 1580–1680*, (London 1982).
52. M. Keen, *English Society in the Later Middle Ages*, (Harmondsworth 1990), p. 76.

6 The Basis of Improvement

There were compensations for the smallholder, not so much in the numbers that were protected from expropriation, for it was usually the market rather than the courts that made the difference between failure and survival, and protection from the market was limited. The real compensatory factor was that the process of commercialization and consolidation of holdings might bring more general economic benefits, offering subsistence to those who were losing their independent position in rural society and becoming labourers. Although wage rates tended to fall during the period, the threat of famine became gradually less severe, in part because employees were often fed at work.

We have seen that the pamphleteer Joseph Lee could confidently claim that it was only the high yield of corn from enclosed fields that explained the plentiful harvests of the mid-1650s.[1] He was in no doubt that enclosure and its associated practices were conducive not only to increased profits but also to new and more productively efficient methods of farming. Traditionally, historians have also assumed that enclosure was the necessary framework for applying new techniques, though a more recent tendency has questioned whether it was really so essential. Robert Allen's view that the open fields were capable of achieving almost equally high yields by the end of the eighteenth century has already been cited.[2] But by that time, common field farming had probably departed some way from its original character and had become in many respects a more commercial activity. In its earlier manifestation, it had tended, by definition, to follow customary lines and was not naturally innovative. Enclosure, on the other hand, was by its very nature a means of applying new methods and increasing productivity, which made it especially important in the sixteenth century, when improvement was in its infancy.

This was certainly the view held by the agricultural experts who assessed the matter at the time. It was substantiated not only by observation but also by their understanding of the degree to which enclosure could raise the value of land, from John Norden's estimate that "one acre enclosed is worth one and a half in common" to Henry Best's report of a threefold increase.[3] These men were monitoring matters more closely than might have been the case in previous periods. The sixteenth and early seventeenth centuries saw the emergence of a body of literature analyzing the basis of agricultural method and promoting maximum efficiency. The writers invariably viewed enclosure and consolidation as the principal, if not the sole basis of improving techniques and profitability. The new literature is thought to represent a qualitative change in the approach to farming, reflecting the activities of a new breed of "ingenious husbandman who was willing to experiment. It was taken for granted that he would be master of his land, that is, that his arable and his pasture would be wholly or largely enclosed".[4]

The first of the farming guides to appear, Fitzherbert's *Book of Husbandry*, was written during the most expansive period of cloth manufacture and sheep farming and was an indication that the intensification of marketing was encouraging the rise of an agricultural science. It is an impressively wide-ranging yet detailed exposition of every kind of farming technique or requirement. It laid down the important principle that arable and pasture be combined "and not the one without the other".[5] But sheep were "the most profitablest cattle a man can have", so when it came to pasture, "therefore I propose to speak first of sheep".[6] Fitzherbert's recommendations in this respect were powerful and persuasive, and all suggested that the crucial advantage was enclosed pasture. He believed that the sheep would thereby avoid diseases like scab, which developed when they were folded in the common field. Enclosure also protected them against storms: "[I]f any ill weather come they will rise up and go to the hedge. And this manner of folding shall breed no mathes nor scab, nor injure their flesh, and will be a great safeguard to the sheep for rotting".[7] He stipulated that those with only common or rough pasture should delay their time for putting the ram to the ewe, but those with "several and sound pasture" could control breeding and wean lambs earlier and more frequently.[8] Accordingly, he concluded by describing the best way to set hedges, with an assurance that this provision would soon pay for itself many times over.

The next fully fledged farming guidebook to be printed in English was the work of Thomas Tusser, thirty years later. Tusser set down a long list of sound farming precepts in a hundred verses. It went through many editions and became a popular textbook among English gentlemen farmers. Like Fitzherbert, Tusser recognized the crucial importance of pastoral farming and was happy to regard it as a sign of good husbandry, which was in turn synonymous with high profits.

> While gentles use walking with hawks in their hands,
> Good husbands with grazing do purchase their lands.[9]

So the obstacles to improvement were not just the crusty, hide-bound smallholders but also the gentlemen preoccupied with hunting. There were more productive activities now available. Tusser believed that sheep grazing enabled the farmer to maximize the value and the extent of his lands. And like

Fitzherbert, he thought that the best assurance of profit and improvement was enclosure.

> Then hedge them and ditch them, bestow thereon pence,
> for meadow and corn craveth ever good fence.[10]

G.E. Fussell thought that Tusser had arranged his book in calendar form "and written in doggerel verse, probably with the idea that rhyme would be of assistance to the supposedly sluggish brains of the rural community".[11] Fussell assumed quite a wide gulf between the new agricultural scientists and the old custom-bound ways. He noted of the first guide to the growing of hops, published by Reginald Scot in 1574, that it was "the work of a practical man, written for practical men, and in this respect is far in advance of most of Scot's contemporaries, who were still much interested in the superstitions of the time, and the pseudo-science of the middle ages".[12] The spokesmen for "improvement" at the time also tended to take a disparaging view of the traditionalist smallholder. Walter Blith, author of *The English Improver Improved* of 1652, depicted them as "mouldy old leavened husbandmen, who themselves and their forefathers have been accustomed to take such a course of husbandry as they will practice, and no other, their resolution is so fixed no issues or events shall change them". In Wiltshire, John Aubrey found that the larger landowners were enterprising, but the husbandmen were "very late or unwilling to learn or be brought to new improvements".[13]

What weight should be placed on this distinction between the emerging agriculturalist embracing the lessons of science and the old-fashioned smallholder adhering to customary usage and seeking spiritual support? The traditionalists should not be seen as "backward". In a sense, the contrast was just between different imperatives, and alternative ways of perceiving the world. The improvers had happened upon the modern mindset, which thinks in terms of individual objects and studies their particular character and effects. But the medieval mind thought in terms of universals and established the nature of a thing not by discovering its structure but by looking up to heaven, "where it exists as an idea".[14] To this extent, the farmers' propensity to be guided by "superstition" simply reflected the inclination of the age to see everything in an undifferentiated continuum and to "ascribe a kind of substantiality to abstract concepts".[15] It was natural for the medieval villagers to invoke spiritual forces rather than technical change to improve their chances in the fields.

Medieval universalism had a good deal to recommend it, in terms of both a properly balanced concept of village society and a form of integrated, sympathetic contact with the land. It represented a different and in some ways more viable way of relating to the natural environment. As suggested above, the medieval farmers occupied the land more by their collective physical presence than by the structures of ownership and control. This left the land free but, by the same token, limited the farmers' conscious command of what the earth might be made to produce. The age was not naturally inclined to the kind of intervention that led to "improvement". Postan observed that, "the most distinctive feature of medieval technology was its immobility". He thought that their knowledge was limited and that what they knew was not always applied. The constraints of communal farming and "a lack of adventurous individualism probably played a part".[16] Thus, without technological development, population would eventually outrun productivity, and the means of adjustment would be a rising death rate. This was what occurred at the beginning of the fourteenth century, when yields were reduced to a low point of six to eight bushels an acre, due to the use of poor-quality seed, a shortage of animals and manure, and a lack of under-drainage, all aggravated by extended periods of bad weather.

Yet the medieval farmer should not be burdened with the charge of "pseudo-science". In the mindset of the time, there could be something positive in *not* taking a "scientific" approach. It was neither desirable nor advisable to interfere too much with what God had provided or intended. God gave to all things their nature, and it might even be regarded as sacrilegious to presume to "improve" it. In fact, in the Middle Ages, improvement was not a technical but a moral concept, essentially about becoming better by the example of Christ. So it was the general inclination of the medieval world to respect things as they appeared to be given and in that sense to leave them as they were. But by the middle of the seventeenth century, these inhibitions were losing their power, and Joseph Lee could specifically reject the restrictions and endorse the practice of the manipulation of natural forces that was the basic dynamic of improvement.[17]

The new agricultural "scientist" was establishing a different relation to the physical world, encouraged from various directions. Easiest to overlook is the fact that market opportunity had focused minds on the possibilities of practical and individual provisions. More generally, the Renaissance, in most of its many aspects, had engendered a new spirit of enquiry, which, in contrast with the medieval way, *did* look into the structure of things. And we have seen that the Reformation too had involved a more rationalist turn of mind. Previous chapters have described this as a rebalancing of the relationship between the physical and intellectual fields. In the medieval world, all things had operated interchangeably in the same plane. But now the intellect was being separated out and ascribed its own inviolable integrity and distinctive conceptual force. By the same token, the physical was being set apart and perceived as something more controllable.

What this constituted in effect was a change in the definition of reality. The prevailing view in the medieval world had defined realism as the belief that the universal idea *was* the reality, which imposed its own conditions. This was being supplanted by the approach more familiar to us today that treats reality in terms of discrete particulars that are there to be examined and managed. The new agriculturalist could no more control the elements than his medieval forbear, but by, in a sense, standing back and observing thc land in a more detached manner, as a field of activity and study that could be subject to "improvement", he could attain levels of productivity and profit which meant that, even at the worst of times, he would emerge as the net gainer.

In the early examples of farming and surveying books, produced in an agrarian world that was still largely medieval in shape, we can visualize the open field context and recognize how it was conducive to the unbroken physical exercise of customary and communal ways. But as the literature developed, it illustrated the emergence of the more individualist forms of enclosure and consolidation that facilitated the process of detached assessment and paved the way for "improvement".

It was Fitzherbert who inaugurated the second important genre of land study, with a book of surveying, which appeared at the same time as his book of husbandry. The economic climate of the sixteenth century encouraged landlords to have their lands surveyed, and in that respect too the process has obvious associations with the advent of a more consolidated view of property. The formal purpose was to clarify the disposition of tenants' rights and dues on the manor and to ensure that the maximum rental potential was realized. But the underlying motive was to indicate how to raise productivity and profits. In the context of medieval landholding arrangements, simply getting the shape of the manor down on paper was an important step towards a more rationalized concept of land. And the survey books give other interesting indications of the various ways in which the system of individual holdings was emerging, in distinction to the old collective fabric.

Fitzherbert began by setting out a method of visually identifying the strip holdings on an open field system. This in itself tended to demonstrate how the old pattern of intermingled strips was compatible with a concept of collective occupation rather than individual ownership. The disposition of holdings could only be traced by beginning at one side of the field and recording how the neighbouring strips "butted and bounded" one another, that is, by identifying the farmers whose strips lay next to them. "Then to begin at the east side next to the highway . . . the parson has two lands, the lord three lands, L.B one land, E.G. two lands, the prior two lands, the parson one land, R.X. two lands, the prior three lands, the lord two lands, L.H. one land, the parson one land".[18] As already noted, the word "land" at this time described a working unit—the most that could be ploughed in a single exercise, and it became in effect the basic, discrete area of ploughing between the drainage furrows in each strip. So what the butting and bounding produced was not really a survey of "land" but a simple enumeration of "lands" and the identification of which farmer worked each one. No systematic geometrical assessment of an individual holding could be made, for each farmer occupied a large number of lands, widely dispersed across the open fields, and inextricably mixed with those of his neighbours.

At first it seemed as if Fitzherbert was going to leave his assessment of the pattern of holdings there, in the same undifferentiated and unsurveyable state that it had been for centuries past. The plotting of the estate appeared to end without any suggestion that the layout either should or could be changed. Given the open field arrangements, it was indeed difficult to see what could be offered in the way of more formal, mathematical methods of measurement. But then he came to the surveyors' other brief, to show how the value and usage of the land could be enhanced. And his overall recommendation was that by far the greatest improvement that could be made was to consolidate and enclose. "Then let it be known how many acres of arable land every man hath in tillage, and of the same acres in every field to change with his neighbours, and to lay them together, and to make him one several close in every field . . . and also another several close for his portion of the common pasture, and also his portion of the meadow in a several close by itself".[19] He suggested that it should be done by consent and might actually protect the small farmer against encroachment by the more powerful. In reality, it was of much less benefit to the commoners than to the substantial farmers.

Fitzherbert became the benchmark for studies of surveying just as for agricultural guides. But other books indicated further developments in the direction of a more consolidated view of landholding. Valentine Leigh was happy to acknowledge that his book of surveying, published in 1577, followed Fitzherbert's method of "butting and bounding" the holdings in the open fields. But he also wanted to point out that in recent times, new ways had been developed for recording the sum of each farmer's holdings. Whereas previously each strip holding had been recorded separately in the context of each particular field, now the tendency was "in the same entry, all in one, they do particularly butt and bound and also enter every man's arable land, also his meadows, closes, and pastures, belonging to every man of the same several tenements, altogether".[20] By this means it was possible to know at a glance "the parcels of lands that belong to every messuage", whereas before it had been necessary to "seek out in the generality of the whole survey of fields".[21] So the strips, even where they remained fragmented, could nevertheless now be conceived in terms of a single landholder. The most interesting aspect of this is that the change described by Leigh underlines how in the medieval context there had been no perceived need to record a discrete individual holding in any respect.

By the end of the sixteenth century, some further advances had been made in the practice of land surveying, and it was acquiring a recognizably modern form. Surveyors and mapmakers like Christopher Saxton and John Norden pioneered a system of mathematical measurement, using chains and plain tables to determine area and measuring the angles on the ground by means of the recently invented theodolite, a graduated circular plate, with an index of sights, and a needle that always pointed north.[22] In a sense, these developments completed the change in the relationship between people and the land.

In the medieval context, the fields had been occupied essentially by the working presence of the farmers, and the disposition of holdings was conceived largely on the basis of communal custom and memory. Now, the land was becoming the subject of precise mathematical and individual assessment.

But it was only really possible to apply scientific methods where consolidated individual holdings were already emerging. This becomes clear in Norden's work. Like the agriculturalists, he recommended enclosure as a means of improvement. "I think it the most beneficial course the tenants can make to increase their abilities".[23] But unlike Fitzherbert, Norden seemed to face no problem of "butting and bounding" the strips in the open fields. There seemed to be no requirement for it, at least not on the estate in question, and he took no steps to illustrate it. The exercise of butting and bounding does not figure in his plotting of the manor. He simply takes up certain stations around the estate and records measurements from them with the theodolite and plain tables, sometimes confirmed by the use of the chain tape. "I hold it most fit to begin about the middle of the manor, and then to take a course as the convenient lying of land will move us".[24] He comes to a common, which he measures in the same way, but there is no mention of strips in open fields. There is regular reference to closes: "[H]ere are some few closes more, and then at an end".[25]

It seems that the evolution of the science of surveying and the consolidation and enclosure of farms went hand in hand. Probably the greatest debt that Norden owed to Fitzherbert was the determination with which the latter had promoted the practice of enclosure. It has been said that the new breed of surveyor arising in Elizabethan times, "gave the lord an independent way of seeing his estates rather than relying on the self-description of the tenants".[26] But Norden would have found it much more difficult to apply mathematical methods of measurement if the new landholding arrangements had not to a great extent taken shape over the course of the sixteenth and early seventeenth centuries. It is also easy to see how the emergence of more consolidated patterns encouraged a new and more detached way of relating to the physical environment. The land was being pinned down, set at a conceptual distance, captured on the page and assessed in theory, rather than simply worked as a continuous, unbroken physical exercise, "subject only to the hand of the husbandman". Enclosure offered a psychological as well as a practical basis for the process of "improvement". This became explicit in the work of Gervase Markham, who hailed a new age of knowledge, in comparison with which "the knowledge of our forefathers . . . is but mere ignorance". He looked forward to "the conquest of the earth, having conquered nature by altering nature".[27]

Markham's words were not just prophecy. In the preceding decades, there had been strong indications that while not conquering nature, men were indeed beginning to direct it more effectively to their own ends. The middle years of Elizabeth's reign registered an exceptional series of good harvests.

It began to appear that however much enclosure had taken place, it had not adversely affected the productivity of tillage. On the contrary, it seemed that despite the process of enclosure, or perhaps even because of it, the supply of corn was becoming more stable and secure. And so, "Good harvests eroded the memories of dearth and helped to disconnect the link between the survival of society and the need to restrain" the freedoms of the farmers.[28] The government began to feel, for the first time, that there was a case for reconsidering its opposition to enclosure and relaxing its time-honoured determination to ensure the availability of corn by every possible means. The change in such a settled, central policy is powerful evidence that the agriculturalists were correct and that the enclosure and consolidation of farms brought about a significant and generally beneficial increase in agricultural productivity.

There is also a need for a clearer understanding of the time at which this began to take a noticeable effect. It is not sufficiently recognized that there actually *was* a significant improvement in the performance of the farming sector in the Elizabethan age. The recent tendency has been to underestimate it. This stems in part from a general inclination among modern historians to diminish the reputation of Elizabeth's reign. It may also be because Eric Kerridge originally seemed to overstate the case, by talking of an "agricultural revolution" between 1550 and 1650. Historians usually associate economic revolutions with notable inventions and innovations, of the sort that became more apparent after 1650, and Kerridge's critics were inclined to minimize the less conspicuous developments that took place in the earlier period. This led many to overlook the real achievement of Elizabethan farmers. Mark Overton, for instance, makes little of it in his overall assessment, even though his statistics seem to show that there was indeed a strong upward trend in general agricultural productivity between 1550 and 1650, after which it tailed away. And most strikingly, the years from 1590 to 1630 saw a sharper rise in wheat yields than in any period until the 1830s.[29] By the same token, it is difficult to understand why Robert Allen places so much focus on the seventeenth century, when his figures indicate that two-thirds of the overall improvement in corn yields occurred between 1550 and 1650.[30]

These figures tell a very important story, which is reflected in other features of the Elizabethan period, which was, for the greater part of its long duration, a time of remarkable well-being. Charles Palliser has noted the salient indicators. Life expectancy was "exceptionally high" between 1566 and 1621. It peaked in 1581, at a much higher level than elsewhere in Europe. And between 1566 and 1586 "mortality was lower than it was to be again until 1815".[31] The latter date is no surprise, for it is around 1800 that historians normally regard the agricultural revolution as having taken effect. More strikingly, it seems that to a significant extent this had already been anticipated in the favourable conditions of the middle years of Elizabeth's reign. On this basis alone, there can be little doubt that it was Elizabethan England that saw the first fruits of improvement.

There were periods of bad harvests towards the end of the sixteenth century, but the manner in which they were contained tends if anything to confirm the Elizabethan achievement. Judging by the level of prices, the dearth of 1596–1597 ought to have produced the worst subsistence crisis of the century, especially since the population had increased most sharply in the preceding decades. The numbers had risen by a quarter since 1560, and the general populace was theoretically at its most vulnerable. Yet there was no nationwide famine. It has been estimated that just 18% of parishes, mainly in outlying regions of the north, suffered a crisis of mortality.[32] So whereas on past experience this should have been the worst crisis, it was actually much less severe than the famine of the mid-century. The years from 1555 to 1559 constituted the only five-year period in the century when the population rise was temporarily reversed by high mortality. The steady increase in population was not interrupted in the 1590s. It seems that by the end of the Tudor century, the English economy was emerging from the medieval pattern where a rise in population would hit a ceiling of limited agricultural productivity and fall back through the effects of famine. In this sense if no other, it is not inappropriate to conclude that the late Tudor and early Stuart periods witnessed an agricultural revolution.

Is it possible to say more about the specific improvements through which the escape was managed? The reality of agricultural advance in the sixteenth and early seventeenth centuries is sometimes underrated because the term "improvement" was not yet being applied directly to farming practice but rather to the raising of rents and income.[33] But the lagging terminology does not mean that agricultural improvement was not happening, and in any case the best platform for increasing rental income was by raising the potential for profits. In fact, there is every reason to suppose that there were developments of approach and technique taking place throughout the sixteenth century. It is a situation where the general circumstances are as instructive and revealing as the details of new method. The whole logic of economic trends in sixteenth-century England, as well as the changing balance of high advantage between pastoral and arable, entailed a pattern of diversification and complementation, which increased flexibility and intensity of practice. The unprecedented profits of sheep farming, the purposive use of enclosure and the persistent rise in arable prices served to focus minds on the potential of change and produced a decisive break with the traditional habit of passivity. The new approach was applicable to all types of farming, and once the interactions began, changes in market trends could extend the diversity of improvement.

Since grain prices were rising faster than wool prices for most of the second half of the century, it is supposed that there would have been a tendency to capitalize by reconverting land to tillage. Although it is difficult to prove this statistically, historians are justified in assuming that it did occur on a significant scale. The advantages in terms of yield were substantial. The ploughing-in of pasture was an effective way of creating nitrogenous fertilizer, the shortage of which had been the main limiting factor in medieval agriculture. And the gain could be sustained by a pattern of farming that regularly combined pasture and arable, grass and grain. This was what Eric Kerridge identified as convertible husbandry, and it certainly seems to have begun in the sixteenth century. Critics have maintained that it is apparent only in parts of the Midlands. But we shall see ahead that there was a powerful development of regional specialization and exchange in this period, which maximized the effect of such improvements even if they were limited to certain areas. It is interesting to note that the benefits have been estimated at much the same level as those of enclosure. Specializations in particularly favoured items could increase interregional productivity by half as much again. In this context, the restriction of an improvement to selected regions could work to the general advantage.

It can fairly be assumed that one of the benefits of the increase in sheep farming was that it opened up the way to more flexible land use. We should not expect that the application of these methods would always appear as a formal innovation. But no doubt many farmers were, in their own ways, following the principal advice of the agriculturalists, which was not only to enclose your fields but also to combine pasture and arable. "And not the one without the other", said Fitzherbert firmly. He was certainly aware of the advantage of using manure in a more focused fashion and understood that enclosure would raise the value of land by half as much again, "by reason of the composting and dunging of the cattle that shall go and lie upon it both day and night".[34] By integrating the two, it was possible to increase stock density, which was a good indicator of improvement. More animals and more manure corrected the shortage of dung that was another major problem in medieval times. This allowed an extension of the arable and generated higher yields. Here was an important part of the secret. The widespread conversion to pasture that had taken place in the first half of the century had led to a pattern of diversity and interaction, which had enhanced the agricultural product overall.

Diversification thus involved many specific aids to fertility. And there was one additional source of improvement, one particular fertilizer that was probably the most crucial. It has not been much emphasized by historians, though it was needless to say, mentioned by Fitzherbert. This was the more discerning and increased use of lime. It may not seem like a sufficiently structured advance to be at the heart of a revolution, but that was probably its power. The vital effect of lime was that, in reducing the acidity of the soil, it freed the nitrogenous nutrients to be taken up. This again addressed the shortage of nitrogen fertilizers that had been the main technical handicap of medieval agriculture. Tudor farmers may not have understood it at a scientific level, but by the later sixteenth century the practical benefits were clear enough for the exceptional value of the practice to be recognized. Thus in Dorset, the process of enclosure "was accompanied by other improvements in land use. Lime was liberally used as a fertilizer".[35] The same was true in

Somerset, Wiltshire and Gloucestershire. And in Devon and Cornwall, the process of enclosure was followed by the wider use of lime and other fertilizers.[36] Liming acid soils in Devon appears to have begun in the last quarter of the sixteenth century. The change in Cornwall was especially striking because by the end of Elizabeth's reign, a region that had once struggled to produce enough corn to feed itself was now in a position to export the commodity. Liming was also noted as used to great effect by "good husbands" in Wales.[37] In early seventeenth-century Sussex, Sir Thomas Pelham "strove to maximise his returns" by denshiring and ploughing in areas of grass. "Lime was spread at the same time".[38] Liming must be regarded as a genuine and highly significant innovation of the age. It was recognized as such by contemporaries and was the factor most often mentioned in explaining the rise in yields. In 1616, it was noted that "the husbandry of liming the ground for corn was first practiced within the memory" of living generations.[39] In 1578, a contemporary commentator summed up the general effect thus: "[I]n later years since the knowledge and use of liming was found there groweth more plenty of corn".[40]

The benefits were sufficient to warrant a great deal of extra effort and expense to produce the lime. In 1618, John Norden noted that "in Shropshire, Denbighshire, Flintshire and now lately in some parts of Sussex, the industrious people are at a more extraordinary charge and toil. For the poor husbandmen and farmers do buy, dig and fetch the limestone, two, three, four miles off, and in their fields build limekilns, burn it and cast it on their fields, to their great advantage".[41] The willingness to engage in this laborious and expensive procedure confirms that a substantial increase in productivity and profits was expected from it. It is probable that the use of lime was indeed the most effective and important means of improving the arable yield. And it seems that when the gain was so clear, "the poor husbandmen" would also embrace the way of improvement.

The basis for increased yields in the high Elizabethan period is apparent. But it led to a degree of overconfidence, which was punished by the bad harvests of the mid-1590s. These years of shortages produced the last official assault on enclosure. The government and many MPs felt it their duty to the poor to take aim once more at the most obvious and indeed only available target. The most interesting contribution, however, came from the MP Henry Jackman, a London cloth merchant, who defended enclosure in the name of economic rationality. He argued that the shortage of grain was not caused by enclosure and the reduction of tillage but by the exceptionally bad weather and the resulting bad harvests. He reminded the House that the years up to 1593 had been a time of great plenty, though there had been no more land in tillage than now. He also recalled that in previous times, before the increase of pasture and enclosure, the land had not been free from dearth. He gave an example from Henry III's reign, "when the whole land lay in open fields, subject only to the hand of the husbandman" and yet the price of corn had risen twelve-fold in a year of bad harvest. Furthermore, he noted that the kingdom of France, which he regarded as still almost entirely given to tillage, was nevertheless thought to have experienced worse famine conditions than England during the crisis of the preceding few years.[42]

Jackman's judgement was soon to be borne out. The fact was that, notwithstanding the exceptional focus on pasture and sheep farming and partly even because of it, the kingdom had become effectively self-sufficient in corn by the end of the century. The Tillage Act of 1598, even while responding to the food shortages, nevertheless recognized that the high levels that could now be attained by English arable production was "a cause that the realm doth more stand upon itself, without depending upon foreign countries . . . for bringing in corn in time of scarcity".[43] Aware that an important threshold of production had been reached, the specialist corn producers of the Midlands and East Anglia lobbied to be allowed to freely export their surplus and fretted against the age-old restrictions by which medieval governments had sought to control the export of corn in favour of the home consumer when prices were high. The legislation of 1361, forbidding the export of corn without a royal license, was not repealed until 1624. But from the end of the sixteenth century, the restrictions were eased. Leading economic opinion now held that allowing producers the freedom to export corn at high prices would actually encourage production overall. Camden noted that it was with the prospect of profiting substantially by exports that the farmers, "began to ply their husbandry more diligently than before, yea and above that which the laws afterward made required".[44] Under these inducements, England was becoming a net exporter of corn. At the turn of the century, Francis Bacon could observe that whereas the kingdom had once been "fed by other countries" it was now returning the favour.[45] This signified an important change in the balance of the agricultural economy, if not a revolution.

NOTES

1. Joseph Lee, *A Vindication of a Regulated Enclosure*, (London 1656), pp. 7–9.
2. R.C. Allen, *Enclosure and the Yeoman*, (Oxford 1992), p. 15.
3. J. Norden, *The Surveyor's Dialogue*, (1618), ed. M. Netzloff, (Farnham 2010), p. 90; Henry Best, *Rural Economy in Yorkshire*, (1641), Surtees Society 33, (1857), p. 129.
4. J. Thirsk, *The Agrarian History of England and Wales*, (Cambridge 1967), IV, p. 162.
5. Fitzherbert, *Book of Husbandry*, (1523), ed. W. Skeat, (London 1882), p. 9.
6. Ibid., p. 42.
7. Ibid., p. 28.
8. Ibid., p. 44.
9. Thomas Tusser, *A hundreth good points of husbandry*, (London 1557), BL.C. 31. c. 26, verse 17.
10. Ibid., verse 68.
11. G.E. Fussell, *The Old English Farming Books*, (London 1947), p. 8.
12. Ibid., p.12.

13. Blith and Aubrey, cited in K. Wrightson, *English Society 1580–1680*, (London 1986), p. 136.
14. J. Huizinga, *The Waning of the Middle Ages*, (Harmondsworth 1976), p. 206.
15. Ibid., p. 210.
16. M.M. Postan, *Medieval Economy and Society*, (London 1972), p. 41.
17. Joseph Lee, *A Vindication of a Regulated Enclosure*, pp. 20–1.
18. Fitzherbert, *Surveying*, (London 1523), in *Certain Ancient Tracts*, ed. R. Vansittart, (London 1762), BL 1508/1096, p. 70.
19. Ibid., p. 96.
20. Valentine Leigh, *The most profitable and commendable science of surveying*, (London 1577), BL 967. k. 15. unpaginated but as counted, p. 61.
21. Ibid., 63.
22. C. Palliser, *The Age of Elizabeth*, (London 1983), p. 199.
23. John Norden, *The Surveyor's Dialogue*, (1618), ed. M. Netzloff, (Farnham 2010), p. 90.
24. Ibid., p. 113.
25. Ibid., p. 124.
26. R. Hoyle, ed., *Custom, Improvement and Landscape*, (Farnham 2011), p. 13
27. Gervase Markham, *Markham's farewell to husbandry*, (London 1620), pp. 3, 9.
28. J. Appleby, *Economic Thought and Ideology*, (Princeton 1967), p. 57.
29. M. Overton, *The Agricultural Revolution in England*, (Cambridge 1996), p. 87.
30. R.C. Allen, *Enclosure and the Yeoman*, p. 207.
31. C. Palliser, *The Age of Elizabeth*, p. 46.
32. M. Overton, *The Agricultural Revolution*, p. 131. For the basis on which the problem was contained, see M.J. Power, "London and the Control of the "Crisis" of the 1590s", *History 70*, (1985).
33. P. Warde, "The idea of Improvement", in *Custom, Improvement and the Landscape*, pp. 132–3.
34. Fitzherbert, *Book of Husbandry*, pp. 9, 22.
35. J. Thirsk, "The Farming Regions of England", *Agrarian History of England and Wales*, IV, p. 67.
36. Ibid., p. 75.
37. Ibid., pp. 117–8.
38. A. Fletcher, *A County Community in Peace and War*, (London 1975), p. 15.
39. Ibid., pp. 133–4.
40. C. Palliser, *The Age of Elizabeth*, p. 197.
41. J. Norden, *The Surveyor's Dialogue*, p. 180.
42. J.E. Neale, *Elizabeth and Her Parliaments*, (London 1957), II, p. 342.
43. *Tudor Economic Documents*, ed. R.H. Tawney and E. Power, (London 1924), 1, p. 84.
44. C. Palliser, *The Age of Elizabeth*, p. 193.
45. Ibid.

7 The Changing Face of the Land and the "Great Bravery of Building Which Marvellously Beautified the Realm"

There was undoubtedly a revolution in the physical environment. By the second half of the sixteenth century, it was greatly altered from the shape in which it had lain throughout the medieval period. The most obvious change was the disappearance of the ecclesiastical houses, which had made up the ever present and all-encompassing face of the universal church. They had been the most imposing features of the countryside. These were the buildings that, above all others, had defined the character and cultural balance of medieval life. Their importance was accentuated because, unlike elsewhere in medieval Europe, there were no substantial towns in England except the capital. The castles had been few and far between and always outside the normal tenor of life. The villages, though many, had been physically insignificant, tiny low huddles, marked only by the spires of their churches. But all around there were the abbeys, priories, chantries and chapels. It is not a great oversimplification to describe the medieval landscape as one of open country punctuated by ecclesiastical sites and structures. Then suddenly, the ubiquitous physical form of the independent church was no more. The ecclesiastical houses were reduced to rubble within the space of a few years. It was as if some geological shock wave had destroyed an entire civilization.

The rich patterns of communal and devotional mobility that had revolved around the universal church and its shrines were also disappearing from the scene. The habits of procession and collective presence had been the dynamic of agrarian work and society—the manifestation of the indivisible continuum of medieval life, in which the physical and the spiritual were combined interchangeably in one undifferentiated notion of reality. Now those unbroken flows of movement were seeping away. The spring and summer of 1547 saw the last performance of the endless play of ceremony and celebration through which the people of the villages bound heaven and earth together.[1] The rite of pilgrimage, so long the necessary way to spiritual grace and worldly success, was terminated, and the working company of the Saints withdrawn. The act of worship retired into the preaching house of the church, and the devotional life of the open countryside came to an end.

The changes in the shape of landholding and land use reinforced the effect. As the processional and communal round slipped away, so the hedges, ditches and fences were slotted into the space. Whatever the exact amount of enclosure that had occurred by this time, it was on a sufficient scale to be noticed as a reordering of the landscape and a disruption of the few aspects of ritual that survived into Elizabeth's reign. The frequency with which the Rogation week perambulations were obstructed by the erection of hedges and fences or the obliteration of ancient routes and paths is a good indication of the degree to which the countryside was being reshaped, and the time-honoured context of communal activity and mobility was being displaced.[2] The old, open perspectives were disappearing. Henry Jackman denied that enclosure was detrimental to the production of corn but did not attempt to hide the fact that it was changing the agrarian scene and creating a prospect perceptibly different from that of medieval times, when tillage had been the dominant feature and "the whole land lay in open fields, subject only to the hand of the husbandman".[3] He contrasted the Elizabethan landscape with that of France, which seemed to have retained the characteristics of the traditional economy of the Middle Ages, dominated by open tillage, with a sense of the fields full of people. France had also retained its ecclesiastical houses, along no doubt with the accompanying habits of pilgrimage and procession that had also served to fill out the countryside. In England, many of the purposes for which people had occupied the fields had now been discontinued. The sheep had not actually eaten the people, but in parts of the country it certainly looked as though they had.

The changing face of things in England did not go unremarked by others. More's famous visitor from Utopia was not the only one to perceive a rural context dominated by sheep. In Dorset, the most notable feature of the sixteenth-century landscape, recorded by John Leland and all contemporary observers, was the vast flocks of sheep that now occupied a great part of the county. In the early sixteenth century, local communities brought suits in Star Chamber against local landowners enclosing the commons, but it failed to stop the tide—the area of Dorset devoted to sheep farming was greatly increased.[4] On the other side of the kingdom, East Anglia also drew attention for the advent of immense flocks of sheep.

Sheep and enclosures were the principal features of a visible change in the balance of agrarian life, which was losing the mainly arable aspect that had prevailed in earlier centuries in England and still obtained elsewhere in Europe. Foreign visitors were unanimous in noting the distinction. Nicander Nucius, in 1545, saw "so many flocks of sheep that wonder arises in the beholders on account of the multitude of them".[5] In 1592, Frederick, Duke of Wurttemberg, did not write much about the landscape, but in travelling through the area between London and Oxford, he could not help but observe that "such immense numbers of sheep are bred on it round about that it is astonishing".[6] Estienne Perlin, writing in 1558, noted the distinctive structure: "[T]he fields are all enclosed with hedges, oak trees, and several other sorts of trees, to such an extent that in travelling you think you are in a continual wood".[7] This encapsulates the impression that the

English countryside had acquired a different, more closed-in aspect from the traditional medieval scene, which had been characterized by open tillage. All these foreign observers were agreed that England now devoted a much greater proportion of its land to pasture than did other European kingdoms. The more subtle change, less easy to discern, was that this had often gone hand in hand with an increase in the production of corn. People were used to seeing the wheat and barley growing free across the open fields. Who would suspect that it might be growing more plentifully behind the hedges?

One of the most interesting assessments is *The Debate Between the Heralds*, a comparison of England and France, as seen by John Coke in 1549. The French herald proclaims that "France is a world of people; in so much that I think there be more labourers of vines in France than people in England in all estates". In this sense again France reflected the traditional medieval landscape, composed in the main of cornfields and vineyards and full of the village farmers and their families going about their work. It is striking that the large kingdom of France could be perceived as full of people, whereas in the smaller kingdom of England, "it is otherwise, for a great part of it is waste . . . not inhabited or the earth tilled, but consisteth in forests, chases, parks and enclosures . . . many gentlemen for their private commodities enclose a mile or two about their houses, destroying thereby . . . farms and villages".[8]

The distinction may be overdrawn, but it reflects an important fact. The differing impressions were not simply about the numbers in the countryside but also about the manner in which the fields were occupied. In the common field areas of Northern France, there had been no systematic enclosure, so the flow of movement through an open arable land had been sustained, and the broad presence of the village farmers still typified the scene. But in England, significant areas of farming country now presented a contrasting impression of a land emptied of people and closed in with trees and hedges. The English herald confirms the essential truth of this but makes the interesting point that the enclosures could be returned to tillage if required. This both underlines the fact that an exceptional amount of enclosure had taken place and indicates one of the reasons that it proved less detrimental to the performance of tillage than might have been anticipated in traditional terms.

The physical environment was fundamentally altered, in a variety of ways, which were often connected. It is useful to ascribe some kind of overall character to the change that was taking place. It may clarify our perspective on the medieval world. The period tends to be seen in terms of its restrictions, such as the limited horizons of economic and cultural affairs and the absence of liberal freedoms. But it may be more instructive to note the many aspects of medieval life that displayed a special quality of openness. The distinctive character of the disappearing medieval scene consisted in a context of undivided and undifferentiated relationships. That world was *open* in the communal working habits of the farmers, in the dispersed but interlaced pattern of landholdings and in the lighter touch upon the land, just as it was open to the reception of direct spiritual influences on agricultural activity. This was now being supplanted by a world that operated on a specifically different set of assumptions and that in every respect was inclined to divide things up and close them in, as distinct entities. The faith was now regarded as occupying a sphere that that was essentially detached from the corporeal. By the same token, there was an increasing tendency to separate out physical things as objects of rationalized control and to conceive and use the land, as well as its resources, on that basis. And with the new focus on particular forms went a growing sense of absolute individual ownership and command. So once again the process of enclosure was significant beyond its mere formal incidence. Enclosure was not just part of the change but also its sign and symbol.

The Great Bravery of Building which Marvellously Beautified the Realm

There were significant alterations in the form of domestic architecture, which reflected the emergence of a newly self-confident laity and constituted another notable transformation in the physical environment. Now that the gentry possessed the property of the church, their cultural advance was not long delayed, and the architectural dominance of the church came to be superseded by new concepts of secular style. The medieval world had put most of its artistic energies into its ecclesiastical buildings, which had thus commanded the leading public aesthetic, as was indeed appropriate in a society that was conceived *as* a church. But the balance was changing in spectacular fashion.

The only significant secular rivals to the church buildings in the high medieval period had been the castles. They too were a focus of authority in the medieval countryside. But they represented a crude military power, and in both form and purpose they tended to defy the concept of aesthetic style. The castles were, in any case, a declining force. By 1400, the age of the castle as a military command centre was over. Their demise was more gradual than that of their clerical counterparts, but it was equally decisive. The concept of the castle had for some time been ceasing to function as a practical proposition. In this sense too, the fall of the castles was different from that of the monasteries, which were at something of a physical and economic peak when they were wilfully destroyed. Conversely, the castles were undergoing a natural deterioration and were allowed to slip away with rather more dignity. Nevertheless, over the course of many decades, many of them had reached the same state of ruin.

The collapse of the castle as a working concept was recorded by John Leland during his tour of the kingdom in the 1530s and 1540s. Time and tide had dealt severely with the once impregnable fortresses.

> Again and again, Leland tells of castles which are ruinous. Some are mere shells . . . some are become mere heaps of building material; of some the very site is hard to trace. The great castle of Richmond in

> Yorkshire is “in mere ruin”. Chartley in Staffordshire is “in ruin” and “a good flite shot” from the “goodly manor place” which is now in use. The interior buildings of Rockingham in Nottinghamshire are roofless and crumbling, and within the ruins of the neighbouring castle of Berengarius Moyne is now “a meane house for a fermar”.[9]

Many castles had been abandoned as no longer viable and left to their fate. Those that survived into Elizabeth’s reign were no longer being fortified. Some were modernized to emulate the more comfortable quarters of the newly fashionable country houses. And some owners, like Sir John Perot at Carew or the Earl of Worcester at Raglan, had conclusively destroyed the military credibility if their castles by installing huge mullioned windows. The exception, as in other things, was Robert, Earl of Leicester, who even while modernizing Kenilworth Castle, attempted to consolidate its fortifications. At Kenilworth “defence for the last time took precedence over comfort . . . surely the last castle of an over-mighty subject in England”.[10]

By the middle of the sixteenth century then, the core structural fabric of the medieval world had disappeared, either crumbling into the dust or rudely torn down. But rising from the rubble, sometimes quite literally, was a rich pattern of secular architecture. The typical medieval face of the land had vanished, and a new one was taking its place. The combination of the “goodly” manor houses of the gentry and the actually far from “mean” houses of the successful farmers was emerging as the dominant motif in the villages and the countryside. The difference in character was very striking. It was as if it had been necessary for one concept of living to vacate the land before another, with different conditions and requirements, could emerge to replace it.

The aspiration towards a distinctively secular style of building, and the economic basis for it, began to appear in the late fifteenth century. After 1450, there was a noticeable increase in house building by the upper classes, facilitated by the more extensive use of bricks. The trend sometimes featured castellated rooflines, as a halfway house between the traditional military role of the gentry and their increasing civilian status. At the same time, the underlying economic platform for the advance of secular forms can be seen in the impressively framed “Wealden” houses of the successful farmers and clothiers of Kent and Sussex. They still referred to the typical medieval structure, conceived around a central hall open to the roof, but often added a jetty above the service quarters at either end, forming a first floor and embryo wings. By the close of the fifteenth century, this was being developed into a structure framed as two stories throughout, with the hall floored over. This arrangement would become more widely adopted after the middle of the sixteenth century, as integral to the “Great Rebuilding” of farmhouses.

Although the vernacular buildings of the countryside and the late medieval town houses possessed great character and charm, they did not display a coordinated, distinguishable style in the same sense as the ecclesiastical houses. At the beginning of the sixteenth century, it was still the church that dominated the scene, and the architectural quality of church buildings that caught the eye of observers. But the secular forms emerging in the background reflected the increased socio-economic substance of lay society, which would provide a solid basis for the challenge to the position of the church in the 1530s. This in turn would clear the way for a fulfilment of secular style. It was a political as well as an aesthetic concept. The conspicuous presence and prominence of the laity were facilitated not merely by the physical scope created by the disappearance of the monasteries and the concentration of wealth and property in yeoman and gentry hands, but also by the place that these rising classes held in the judicial and administrative structures and the public dimensions of the sovereign state that emerged during the Reformation period.

The aspect of the new secular style that was most apparent to contemporaries was what W. G. Hoskins christened the Great Rebuilding. By this he sought to highlight the process by which from the 1570s onward, the houses of the smaller gentry and the yeoman farmers were being notably extended and enhanced.[11] The economic success of the chief farmers has figured prominently in the foregoing pages, and the extension and refurbishment of their houses were the most conspicuous forms in which they used and displayed their new affluence. Their rise was already under way in the second half of the fifteenth century, but it was in the 1570s that the access of wealth and the trend to architectural improvement were becoming so clear that they amounted to a manifest transformation.

Structurally, the changes usually involved the addition of an extra floor and extra rooms. This meant that the central hall or communal living area was no longer the be-all and end-all of the concept. The hall was divided and surrounded by a larger number of separate rooms. So the typical house now had two floors and an interior that was composed of parlours and chambers. The extra floor also necessitated more attention to the provision of formal staircases and, of course, chimneys.[12] Smoke could no longer be allowed to simply drift out through a hole in the roof of the main living area, as had usually been the case. Now chimneys ushered the smoke invisibly away and at the same time epitomized the higher profile of the houses.

The change was not just "modernization", involving the provision of the kind of internal structure that we would recognize today. It also implied a different relationship with the exterior. In the old concept, the houses sat lower to the ground, and the division from the physical world outside was minimal, especially as there were usually no windows, and the most valuable farm animals often slept within the walls. In the new dispensation, there was greater privacy within, due to the provision of separate rooms, and there was better insulation from the outside, with the introduction of glass windows, curtains and wall hangings. The changes were facilitated by another productive aspect of the Tudor economy, the rapidly expanding manufacture of building materials like brick, tiles and glass.

Contemporary observers regarded the rebuilding process as having begun in the southeast, perhaps the most affluent area, due to the opportunities for farmers to specialize and commercialize their products in response to the ever burgeoning market of the capital. The Elizabethan commentator William Harrison recorded his perceptions of the transformation in his local region of Essex. The multitude of chimneys lately erected was, by its generality and its nature, the most obvious change. The introduction of glazed windows into the houses of the yeomanry was another noticeable addition. These, like chimneys, had once been the preserve of the richer nobility and urban elite. They were now making their first general appearance along the village street. Harrison duly recorded the important implication that there had been a substantial adjustment in the balance between lifestyle and social status. The widespread use of brick and stone in place of timber in the housing of the ordinary independent landholder meant that in this respect he now "rivals what a prince was once". The same could be said of the wider acquisition of costly furnishings and dining vessels, "even unto inferior artificers and many farmers".[13] Harrison's impressions are a good measure of the exceptionally broad base of the consumer society that had arisen by Elizabethan times.

Harrison thought that in some respects the improvements took a while to percolate northward. He noted, for instance, that the institution of feather beds in place of straw pallets had yet to reach Bedfordshire. This was perhaps the cause of some personal disappointment. But in general it seems that the changes were echoed, with quite striking rapidity, far across the country and down the social scale—a reflection of the widespread success of commercial farming and of the socio-economic, political and indeed physical homogeneity of Elizabethan England. William Smith, writing of Cheshire in the 1580s, noted the transformation in the lifestyle of the farmers. They had once "had their fire in the midst of the house against a hob of clay, and their oxen under the same roof. But within these forty years it is altogether altered, so that they had built chimneys and furnished other parts of their houses accordingly".[14] Carew observed the same transition in Cornwall, at much the same time. The houses of the farmers had once had "walls of earth, low thatched roofs, few partitions, no glass windows, and scarcely any chimneys, other than a hole to let out the smoke. But now all these customs are universally banished, and the Cornish husbandman conforms himself with a better supplied civility to the Eastern pattern".[15]

In the Great Rebuilding, the group that had benefited most from the conditions of the Tudor economy, the middling sort of yeoman farmers, announced their arrival as people of socio-economic and local-political substance. And the rise of the yeomanry contributed in turn to the rise of the gentry, certainly in numbers, and doubtless with the injection of wealth. One obvious effect of the increasing substance of the chief farmers was to augment the gentry as a class. The strata of English society were not unbridgeable. The status of gentleman did not rest entirely on inherited dignity. The qualifications were typically pragmatic and, in William Harrison's words, could be met by any man who "can live without manual labour, and thereto is able to bear the port, charge and countenance of a gentleman".[16] Many yeomen were attaining that level of wealth and with it the capacity to make the transition into the ranks of the gentry.

On that basis, together with their own economic success and the elevation of younger sons, the gentry had increased enormously in numbers. Lawrence Stone suggested that over the course of the sixteenth century, they were multiplying about twice as fast as the general population. This was an area of social inflation that the parsimonious Elizabeth was apparently unable to control. She had no direct power over "the unprecedented torrent of claims for arms" during her reign. The overall consequence "was a dramatic increase in the number of the gentry".[17] It is a conclusion confirmed more recently by Mark Overton. "It seems clear the gentry did increase considerably in numbers from the mid-sixteenth century".[18]

One of the most striking illustrations comes from Shropshire, where there were forty-eight gentry families in 1433 and four hundred and seventy in 1623. This helps us to envisage the effect of the proliferation on the ground. The gentry were becoming the unavoidable focal points of social and political power in the countryside. In the sixteenth century "the country-house civilisation was just beginning to emerge, beginning its four hundred year dominance of English society".[19] This signified, among other things, a new way of establishing command over the land, both as a physical and a political space. We have seen how the old nucleated villages were being diminished or dispersed, as the common fields were partitioned and enclosed, and the consolidated estates and farms were set up, often at a distance, around larger, detached houses. The individual power of the squires, together with their ubiquity, gave them the basis of a collective force beyond the locality. They came to represent power over *the land* in both senses of the word that are under discussion here.

There thus emerged what became the classic context of English rural society—each village with its squire, its tenant farmers, its labourers and its enclosed fields. The squires had not always been in such plentiful supply. It has been estimated that at the end of the fifteenth century, perhaps no more than one village in every four or five boasted a resident squire.[20] This situation altered dramatically during the course of the sixteenth century. W. G. Hoskins underlined the fact that at the beginning of that century, the knights and squires were "comparatively rare on the ground". The evidence for the beginning of the reign of Henry VIII indicates a total of about one thousand squires. "But if Thomas Wilson's figure of 16,000 at the end of the century is anywhere near the truth, then the Elizabethan village must have shown an entirely different picture".[21]

So it did, and the difference was clear not only because the houses of the gentry had become an unavoidable presence but also because they had acquired a new and more imposing face. If the Great Rebuilding of the farmhouses was impressive, the new generation of manor house asserted itself at a different level again. It possessed qualities of architectural projection and self-confidence that gentry houses had not previously displayed. There was a transcendent sense of style, where previously there had been very little. The manor houses of the medieval period, especially in the early centuries, were often basic in form, despite their socio-economic status and the fact that they functioned as the administrative and judicial centres of wide areas of country. The manor house might be distinguished by being constructed of stone, like the church, which set it apart from most of the surrounding timber cottages and farmhouses in the village. But the stone was used very crudely and almost entirely without a sense of styling. In their original conception, the manor houses of early medieval times rarely seem to have transcended a heavy, rectangular and unrelieved block-like shape.

The one claim to graciousness in the early medieval manor house was that it was usually conceived around the idea of the lord's hall. This was in fact the only idea that the house incorporated. It was sometimes marked by one large bay window at the end of the hall, behind the judicial dais, emphasizing the semi-divine status and purpose of lordship. In a sense, the building simply *was* the hall, which served as the living quarters and sleeping chamber for the family, as well as the manorial courthouse. The hall was often open to the roof, while the kitchen and servants' quarters would be tacked on behind. Below, there might be an under-croft or basement storehouse. In this too, the houses resembled a basic kind of church, with an open interior and crypt below. Perhaps they also gave the impression that one gets in a church even today—of not really being inside at all. As with the early farmhouses, there was only the most limited sense of division from the outside world, and even in the manor house the most valuable animals were often welcomed within the walls. All in all, these houses reflected the distinctively sustained continuum and openness of medieval life.

As time went by, the medieval manor house took on a less crude and more picturesque aspect but rarely attained a self-possessed distinctiveness of styling. Sections were added, but not so as to produce any kind of matching aesthetic balance. There was more feature and decoration, especially with regard to the windows, but without structural regularity. These houses displayed what has been called a "relaxed informality".[22] They were spontaneous rather than self-conscious. This could endow them with considerable charm, as is apparent at Bradley, from fifteenth-century Devon, with its multi-roofed façade like a row of individual cottages (Figure 7.2). The late medieval manor house presented a friendlier face than its cruder predecessors, and even perhaps than its successors, but it remained stylistically naïve. The fifteenth century sometimes saw the addition of a Great Chamber above the Great Hall. But the houses still gave the impression of being low to the ground and piecemeal in conception, rarely transcending a kind of random functionalism. In terms of what was to follow, what they lacked most of all was a concept of clear-cut, self-contained symmetry.

Although the signs of ambition towards a secular style had been appearing since the later fifteenth century, it was the mid-sixteenth century that saw the crucial qualitative change in the architectural face of the gentry. Camden and Harrison both noted that it was in the mid-sixteenth century that the emergence of a new class of manor houses, conspicuously costly and "remarkable for their elegance", began to appear. Modern assessments have borne out their view. A broad survey of Derbyshire, Shropshire, Essex and Somerset showed that, "the amount of country house building in the fifty years between 1570 and 1620 far exceeded that of any subsequent half century".[23] The expansion was driven in part by the inspiration of a new concept of manor house. Architectural practice had changed markedly in the course of the sixteenth century. At the beginning of the century, it became usual for houses to be actually conceived as a two-storey structure, with all that this entailed for a raised profile. At the same time, the influence of the Renaissance began to take effect. It was reflected to some extent in the use of classical and Italianate detail, but above all in the adoption of symmetry.

The transition was clear and can be illustrated across widely dispersed regions. Devon provides good examples of the medieval form. Dartington Hall, from 1390, was one of the grandest of medieval hall blocks, overawing by solidity rather than style (Figure 7.1, end of chapter). And the variegated face of Bradley, from the 1490s, was the ultimate in late medieval informality (Figure 7.2). But by the early seventeenth century, Devon manor houses were typified by symmetry. On the eastern side of the south coast, Sussex boasted one of the earliest examples of symmetrical aspiration, at Parham, begun in 1577 (Figure 7.4). The style was echoed with striking rapidity in other counties through the 1580s, another sign of the exceptional homogeneity of Elizabethan society. In Derbyshire in the middle of the sixteenth century, distinctly medieval shapes had still been appearing. Thus Hazelbadge Hall, from 1549 was earth-bound and block-like, with barn attached and no pretension to symmetry (Figure 7.3). But by later Elizabethan times, symmetry was the order of the day, as displayed in Barlborough Hall from the 1580s (Figure 7.5) and the most iconic and familiar of Elizabethan houses, Hardwick Hall, from the 1590s. Perfect symmetry can also be seen emerging in Norfolk in the same period, with Heydon Hall from the 1580s (Figure 7.6). In fact, symmetry had become the basis of manor house styling and would remain so through the succeeding centuries. It provided the homes of the gentry with a formal aesthetic character that they had previously lacked. The change has been summed up as a combination of "symmetrical facades inspired by Italy . . . and grid windows in native Perpendicular style".[24] The use of the latter illustrates the supplanting of the architectural dominance of the church. "The building of large Perpendicular churches more or less came to a halt by the 1540s, but the tradition mingled with continental influences to produce a new and vigorous secular architecture".[25]

These houses displayed a sense of perfect balance and completion, with scarcely a hint of openness remaining. The face of the medieval manor house had never transcended a kind of basic functionalism, being built for the immediate purpose, without stylistic cohesion and always somehow close to the ground. But the new manor houses had, with a sudden single flourish, taken wing. They manifested a self-conscious, coordinated sense of styling to which their predecessors had never aspired. Now style was everything. When we look at these houses in their controlled, self-contained symmetry, we are not really aware of the natural, physical world at all.

But to note that the new styling transcended functionalism is not to say that it had no function, for it certainly did. Historians have sometimes talked about the function of these houses in terms of their inner workings and their relationship to the estate. But the function of their outward appearance may have more to tell. What the new style amounted to was a way of imparting complete distinction to a house. Its outward connections were much less immediate and direct than the unbroken link that had existed between the medieval house and the agrarian world. The completed symmetry proclaimed ascendancy and power, but of a different sort from the crude bulk of the castles and the hall blocks. The controlled symmetry of the manor houses was the expression of absolute socio-economic command and overall administrative authority, contained in architectural form.

Most importantly, the style reflected the *national* reach of the increasingly substantial and assertive class of gentry that occupied the houses. The ubiquity of the gentry in the villages had underpinned their emergence as a national force. The self-conscious styling of their houses distinguished them from the mundane and referred to a broader culture of high fashion beyond, with a nod to the ultimate supremacy of classical form and imperial pretension. This was the architectural face of the coordinated social and political prestige that the gentry were attaining. They now commanded the leading public aesthetic that had once been the preserve of the church. The controlled symmetry of the manor houses asserted the growing status of the gentry as a new kind of public authority.

Figure 7.1 Dartington Hall, Devon, 1390. (© Hugo Butterworth.)

Figure 7.2 Bradley Manor, Devon, 1495. (© Hugo Butterworth.)

Figure 7.3 Hazelbadge Hall, Derbyshire, 1549. (© Hugo Butterworth.)

Figure 7.4 Parham, Sussex, 1577. (© Hugo Butterworth.)

Figure 7.5 Barlborough Hall, Derbyshire, 1580s. (© Hugo Butterworth.)

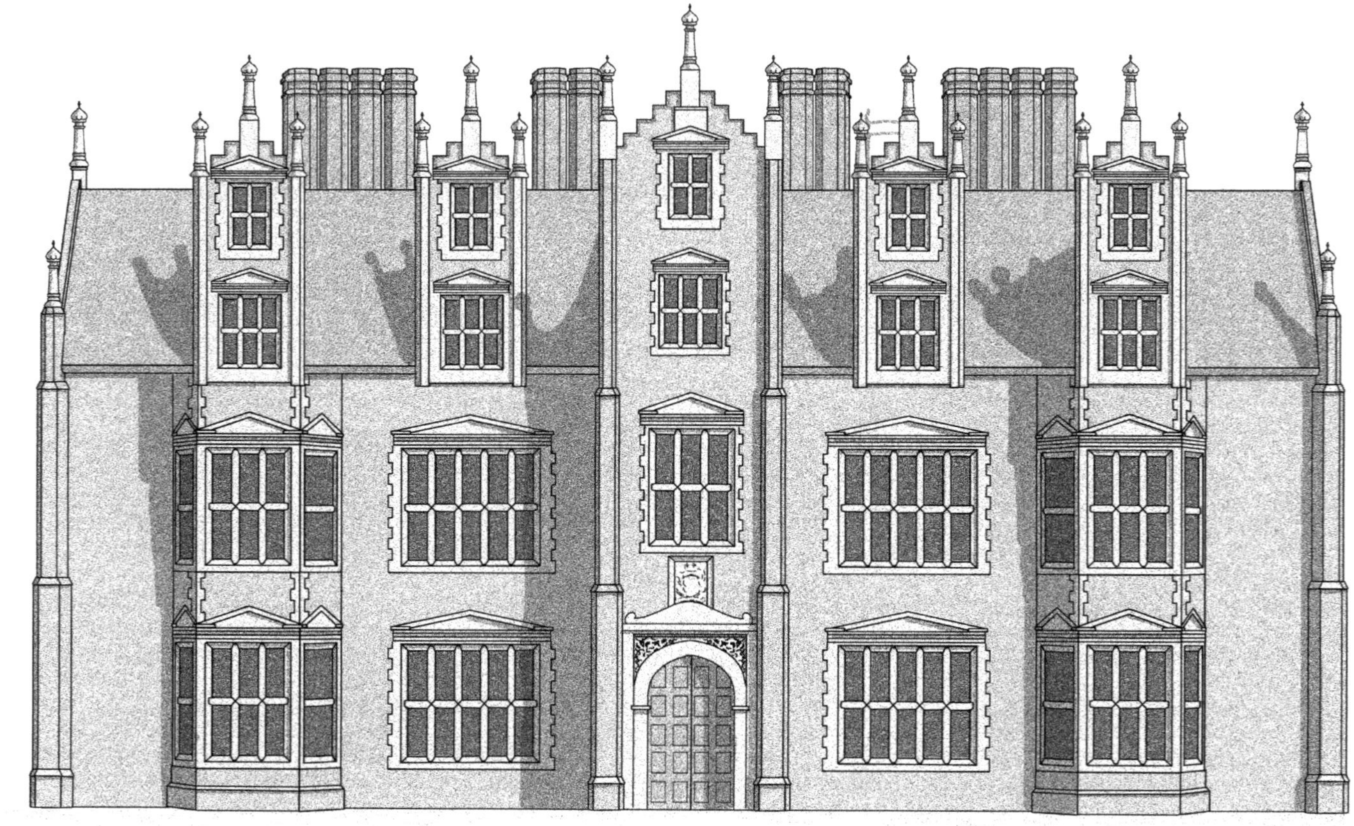

Figure 7.6 Heydon Hall, Norfolk, 1580s. (© Hugo Butterworth.)

NOTES

1. R. Hutton, *The Rise and Fall of Merry England*, (Oxford 1994), p. 76.
2. Ibid., p. 53.
3. J.E. Neale, *Elizabeth I and Her Parliaments*, (London 1954), II, p. 342.
4. C. Taylor, *Dorset*, (London 1970), p. 127.
5. Nicander Nucius, *The Second Book of the Travels of Nicander Nucius of Corcyra*, ed. J.A. Cramer, Camden Society XVII, (1841), p. 18.
6. W.B. Rye, *England as seen by Foreigners*, (London 1865), p. 30.
7. Estienne Perlin, *Description of The Realms of England and Scotland*, ed. R. Gough, (London 1775), p. 25.
8. *Tudor Economic Documents*, ed. R. Tawney and E. Power, (London 1924), III, pp. 1–7.
9. J.D. Mackie, *The Earlier Tudors*, (Oxford 1966), p. 38.
10. C. Palliser, *The Age of Elizabeth*, (London 1983), p. 112.
11. Hoskins placed the process mainly in the period of 1570–1640. The reassessment by R. Machin (*Past and Present* 77, 1977) suggested that the Great Rebuilding continued after the Civil War period but confirmed the starting date of 1570, as well as the peaks of activity in 1570–1589 and 1620–1639.
12. C. Palliser, *The Age of Elizabeth*, p. 110.
13. W, Harrison, *A Description of England*, (1577–1587), in *Tudor Economic Documents*, ed. R. Tawney and E. Power, III, pp. 68–70.
14. Ibid., p. 111.
15. Ibid.
16. *Harrison's Description of England*, ed. F.J. Furnivall, *New Shakespeare Society*, 6 series, I (London 1877), pt. I, chap. 5.
17. L. Stone, *The Crisis of the Aristocracy*, (Oxford 1965), p. 67.
18. M. Overton, *The Agricultural Revolution*, (Cambridge 1996), p. 169.
19. W.G. Hoskins, *The Age of Plunder*, (London 1976), p. 3.
20. J.R. Lander, *Government and Community: England 1450–1509*, (London 1980), pp. 42–3.
21. Hoskins, *The Age of Plunder*, p. 55.
22. J. Steel and M. Wright, *The English House*, (Woodbridge 2007), p. 50.
23. L. Stone, *The Crisis of the Aristocracy*, p. 551.
24. C. Palliser, *The Age of Elizabeth*, p. 369.
25. Ibid., p. 368.

8 The Definition of the State, and the Developing Structures of National Administration

The term "political nation" is somewhat misused and misunderstood. Some historians employ it loosely to suggest a politically active element in the population. Others query it as a vague assertion conveying the idea of a politicized class without adequate substantiation. Some of the critics are historians who dislike the thought of community politicization on principle, but others merely regret the imprecision. The doubts are unfortunate, for they obscure an important development. When properly defined, the term "political nation" could scarcely be more relevant or precise. The clue is in the word "nation", for therein lies the clarity and cogency of the concept. "Political nation" should be taken not as referring to groupings within the kingdom but rather to the framework of a nation in itself. A political *nation* is exactly what was evolving in sixteenth-century England. That is to say, there was coalescing in the body of the kingdom a coherent political structure and self-consciousness with a capacity to assess and express an independent perception of its own unified interests and character.

In the context of this study, the land as just a physical space in the occupation of a particular king or race of people was turning into *a land* with a normative, sociopolitical dimension of its own. It should be emphasized that this took shape as a process of coordination from within the community rather than as an imposed unification from the centre. The emergence of state structures in the early modern period is usually envisaged in the form of the extension of autocratic control by the monarchy. The most conspicuous examples are France and Brandenburg-Prussia, where during the seventeenth century, the crown was able to establish something like "absolute" power by raising arbitrary taxation, mainly by imposition upon the peasantry, through a paid bureaucracy, backed by a powerful standing army. Thus Perry Anderson refers to "the military-bureaucratic apparatus" that was creating the centralized state of Brandenburg-Prussia.[1]

But it is wrong to suppose, as Anderson seems to do, that the early Stuarts were consciously pursuing this European model of absolutism.[2] The distinctive balance of power that had developed in England over the preceding decades meant that military-bureaucratic absolutism was not only impracticable, it could not even be seriously contemplated. The difficulty was not merely that the gentry and their yeoman and merchant collaborators had established an unshakeable grip on the wealth and the military capability of the kingdom. Of more particular significance was the fact that the weight of taxation fell on the gentry and the middling sort, rather than on the peasantry, and for this reason the substantial class of landowners were fully and actively committed to the processes of representative consent. Unlike the nobilities of France and Brandenburg, the English gentry had a powerful vested interest in resisting any move towards arbitrary methods of money raising. This was an almost insuperable obstacle to absolutist ambitions, as was recognized at the time. In the late 1620s, the Privy Council considered various options for discretionary money raising, as employed in other European kingdoms. But as Sir Thomas Edmondes concluded, "it is dangerous to put any of these things in practice, without either the consent of parliament, or the hope of gaining the conformity or submission of the people thereunto".[3] In England, between the late fifteenth and early seventeenth centuries, the representative system had come to embody the coordinated strength of the nation and its ruling class, and it was difficult to see a way around it.

By the same token, the distinctive force and national scope of the representative principle in England meant that the concept of the state interest was appropriated by the gentry and their parliamentary partners, rather than by the crown, which was thereby denied another of the essential foundations of absolutist rule. So in every respect the consolidation of the nation was a popular rather than a royal phenomenon. This did not prove to be an impediment to the development of the state. Few would deny that by the end of the seventeenth century, England had become a coordinated national unit that generated greater economic resources and more effective military power than either of the other kingdoms in view. If the explanation has remained somewhat obscure, it is because many recent historians have been unwilling to accept that there was popular motivation behind the political changes of seventeenth-century England. But in fact there was, and this was also the essential basis of the new nation state.

It has been suggested in the foregoing pages that the medieval period was defined, perhaps above all, by a quality of openness. This applied as much to the political sphere as any other. At the beginning of the sixteenth century, the concept of the discrete, fully independent state did not really exist. The kingdoms of Europe had not yet acquired definitive national shapes. The power of the king or prince was the main focus of political unity, but it was still a movable feast. The authority of the various monarchies was not co-terminus with the areas that we would recognize as the nation states of today. In fact, there was as yet no absolute sense of boundaries. The reach of the crown might extend to theoretical claims in countries beyond the immediate realm. On the other hand, it might be restricted and disputed by regional powers within its own territorial base and by "over-mighty subjects" with their own pretensions to assuming royal authority. These arrangements not only tended to produce disruption in government and disorder in the kingdoms, they also comprised a considerable obstacle to the consolidation of unitary states.

Equally disruptive of full national sovereignty was the position of the other most characteristic force in the medieval world—the universal church. On the eve of the Reformation, the church occupied about a quarter of the land of England. In any given village, the church was likely to be the biggest landholding presence, even if an absentee presence.[4] This reflected the balance of real authority in the kingdom and the fact that the moral power of the church was an active force in economy and society. The physical ubiquity of the church did much to determine the shape of community on the land. And the fact that the greatest landholder was a spiritual presence contributed another dimension to the sense of openness, where the mark of consolidated tenure or public organization was not yet set hard and fast.

The church occupied about a quarter of the kingdom and considerably more than a quarter of the minds of its people. The religiosity of medieval society flowed openly and freely in an uninterrupted continuum from the daily labour of the land to the working concept of the eternal heavens. The farming communities of the open fields and commons reached out indivisibly through the parish churches and through the shrines of the Saints and the ecclesiastical houses to the universal authority of the church of Christendom, and all were as one in the Body of Christ, which was the only complete and singular unity.

The ecclesiastical houses were the physical manifestation of the independent authority of the church, and their presence signified the limits of secular power. Since there were no states as such, we cannot say that the church constituted a state within a state, and indeed to say so would invert the true balance of the relationship, because all secular institutions were corporations within the corporation of the church. The clerical order was an independent legislative authority and embodied the highest form of law—the Law of God. In principle, the secular law also derived from the Law of God. In this sense, the king and his courts did not actually make law—they merely declared the law that God had made.

So at the beginning of the sixteenth century, the scope and power of secular, public authority was limited. It did not yet possess its own discrete sphere. It was specifically restricted not only by regional sectionalism and the continuing force of the private jurisdictions and personal military power of the magnates but also by the universal presence of a moral and spiritual dimension, which more than anything else determined the character of community life. This duality or multiplicity of powers was reflected in the way that men defined and directed their loyalties. Devotional commitment was to the church of Christendom. The feudal ties that characterized social and political affairs were essentially personal, with no other necessary definition. This militated against the development of a single, collective loyalty. In this sense too, the physical land had not yet become a political land.

In this open perspective, loyalties did not focus on a singular statehood. "In the fifteenth century there was no conception of England as a unitary state: regnum and sacerdotium owed allegiance to king and pope respectively".[5] At that time, the word "state" had a social rather than a political connotation. It was used in the sense of "condition" as regards economic status or welfare and could thus also apply to one's condition or place among the orders of medieval society. What it did not yet refer to was a political unit. Yet by the end of the sixteenth century, this change had taken place. As John Guy concludes, "[T]he shift from "realm" to "state" is among the more interesting developments of English history".[6]

Surprisingly, the change is also one of the least analyzed of historical phenomena. We can suggest an outline of some of the causes and implications in relation to the shifting basis of the concept of the common welfare. In the medieval period, the concern for the common good focused on one particular "state", or order, of society. It was essentially the good of the "commons", the most fundamental and much the most numerous class in the medieval world—the husbandmen who worked the open fields and common lands. Their well-being was indeed a universal interest that extended in a non-competitive and undifferentiated fashion across all the kingdoms of Christendom.

It was in this manner that More employed the term "weal-public" in Utopia. Thus he perceived the extension of enclosure and sheep farming as "much annoying the weal-public". "The husbandmen be thrust out of their own . . . out of their known and accustomed houses". He used the word "state" in the old sense of condition or position in the world. The miscreants were those who refused to stay within the bounds of their given place among the orders of society and who were "not content with the present state of their lives".[7] Robert Crowley was equally concerned to maintain the given balance of the social orders and used the word "state" in the same sense. To those who had exceeded their rights and enclosed the commons, he said, "wish that you had contented yourselves with that state wherein your father left you . . . God hath called you to an office so that ye may with good conscience take ye the state that ye be called unto".[8]

Thomas Starkey also employed the word "states" in the sense of the orders of society. But at the same time he was part of an emerging group of "commonwealths-men", who were rejecting the moralistic and contemplative tendencies of medieval times in favour of active intervention to produce the ideal commonwealth. The clearest indication of a developing national outlook was his concern with the balance of trade and the need to adjust it to encourage home production.[9] The rise of the cloth trade to national and international significance had perforce induced a sharper awareness of the trading dimension. This perspective, focusing on the relations between internal and outward economic conditions, was an important basis on which a more unified notion of the common good was formed.

The most incisive and modern sounding of the mid-Tudor economic texts, *A Discourse of the Commonweal*, makes no use of the word "state" in the sense of orders of society. It is concerned with the good of the commons but suggests that this will be best served not by enforcing social balances but rather in the context of a national perspective, reviewing the conditions of trade. Like other commentaries, it is critical of the extent and methods of the enclosure that has taken place in recent decades but suggests that enforcing restrictions is not the solution; it is better to accept the force of the profit motive and accommodate it by adjusting the terms of trade to give tillage the same freedoms and profitability as pasture. The *Discourse* offers a lengthy analysis of the balance of trade and exchange rates, as well as the state of the coinage. The development of trade and the problem of maintaining it again emerge as important factors in creating a national perspective. And when the *Discourse* uses the word "state", it is in a way that helps us to understand the manner in which the meaning of the word was being transposed. The *Discourse* is ready to refer to "the state or necessity of the country".[10] Wealth, or welfare, was increasingly being assessed in a national dimension. And one's "state" or condition in the world was coming to be seen less in terms of a God-given position among the orders of the universal society and more in terms of one's interest in the requirements or "state" of a discrete national community, a single, overall estate, which thereby became the "state" interest.

Piers Ploughman's Exhortation was written at much the same time as the *Discourse*, and although it gives a less balanced and comprehensive picture, it takes a similar line. The enclosure of arable land for sheep is again condemned, but just as in the *Discourse*, the suggested solution is less social than national: essentially that more freedom and profitability should be given to the trade in corn rather than restricting the use of pasture. Piers also offers an interesting early use of the word "state" in its more modern sense, saying that although the displacement of the commons from their customary rights is to be deplored, he does not believe that all land should be held in common, for no state can be governed in that way.[11] The new nation state would come to be associated with a different model of landholding.

The break with Rome had in practice created an autonomous state, with a set of independently definable national interests. It had also established the most important inner dynamic of political unity. Apart from the emphasis on separate trading interests, the other crucial element in the growing sense of a national state was the emergence of an overriding, omni-competent statute law. The 1530s saw the first appearance of a fully sovereign, secular lawmaking power, and it derived its authority specifically from the capacity of parliament to represent the entirety of the kingdom. This developed through a series of statutes by which parliamentary law superseded and in effect destroyed the independent legislative authority of the church.

The new concept of sovereign representative law was given authoritative expression in the 1560s by Sir Thomas Smith (who is also a likely author of the *Discourse*). Again, he makes no use of the word "state" in terms of the orders of society. His groupings appear rather as classes, which is the word that he employs. He treats the term "commonwealth" as synonymous with national government, and he observes that the most important instrument of this is the exercise of the power of legislation in parliament.

> The most high and absolute power of the realm of England is in the parliament . . . and upon deliberation every bill or law made is the prince's and the whole realm's deed, whereupon justly no man can complain . . . for every Englishman is intended to be there present, either in person or in procuration and attornies . . . from the Prince to the lowest person of England. And the consent of the Parliament is taken to be every man's consent.[12]

The idea that the "absolute power" of legislation could only be exercised in parliament carried an important radical potential, for it was a basis on which MPs could assert that other sovereign powers that were once the preserve of the crown should now be performed only by parliamentary consent. In fact, they invoked the theory in precisely this manner in 1610, in their attack on the king's prerogative of setting customs dues by discretion. The crown held discretionary rights in order to exercise responsibility for the general interest, and in this respect we can note a further challenging implication of Smith's analysis. In his description of the nationally coordinated force of sovereign representative law, we can hear the dynamic through which a unitary public sphere could emerge and an independent, popular definition of the common good might be made.

These propositions were further elaborated by Richard Hooker, in the *Laws of Ecclesiastical Polity*, the most important work of political theory to be produced in Elizabethan times. Hooker included a long section on the rules that should guide men's actions but made no mention of an obligation to conform to the bounds of one's condition, or state, in society. He put great faith in the power of "reason". "For the Laws of well-doing are the dictates of Right-reason". And he believed, specifically, that the collective "reason" of the community was expressed in parliament, ultimately through the making of sovereign representative law. Hooker found the phrase that captured the essential basis of this concept—the "entire society". "The lawful power of making laws to command whole politic societies of men belongeth so properly to the same entire societies . . . as in parliaments . . . although we be not personally ourselves present, notwithstanding our assent is by others in our behalf . . . no reason but that it should stand as our deed, no less effectually to bind us than if ourselves had done it in person".[13] As long as that consent had been given, said Hooker, these provisions should be obeyed by every man as "the law of the whole commonwealth or state wherein he liveth". This was the dynamic that bound the new nation together, just as it bound every citizen to obey its laws. Thus it was that the commonwealth became a state, a single estate, as people began to define themselves less in terms of a condition among the orders of medieval society and more in terms of an identity with the special, unifying characteristics of the nation and its position in the world.

Many of these dynamics shared the same structural underpinning, in a distinctive framework of community-based organization that existed in England. This is conveniently identified in the comments of Polydore Vergil, a collector of papal revenues, who arrived in England in 1502. He used the idea of nation in its old sense, as indicating a race of people. Thus he observed that the island of Great Britain was divided into four parts, "whereas the one is inhabited of Englishmen, the other of Scots, the third of Welshmen, the fourth of Cornish people".[14] But at the same time he noted a significant and quite unusual administrative system within the English kingdom, which would in the event provide the basis for a different concept of nation. He recorded that England, the largest of the four parts, "is divided into 39 shires, which commonly men call counties".[15]

It is interesting that Vergil should have been so aware of the existence of the county system of administration, and it implies that English people themselves had become very conscious of it by this time. It merited notice. The way in which the structure had developed and the potential that it carried were exceptional. No other major European kingdom possessed the same indigenous base for a coordinated administrative system—that is, one composed of uniform elements with the capability to exist and function irrespective of regional sectionalism or the private interests of territorial magnates. The regularity and kingdom-wide application of the county network constituted a solid foundation upon which the land of the English would become a land in the modern sense of "nation".

The shire system had been put in place by the West Saxons, as a mechanism for incorporating lesser internal realms as defensive and administrative units, and bringing the whole of England under the authority of one crown. For some centuries thereafter, the county structure was overlaid by the development of feudalism, which sat more naturally with a pattern of private and regional powers. The county framework survived because it gave the monarchy at least the outline of an independent platform in the localities. The counties also came to provide the network of parliamentary constituencies through which the general body of the kingdom could be represented and taxed. But in the exercise of power, it was long overshadowed by the feudal magnates. In the civil disorders of the fifteenth century, the shire reeve, the principal standing agent of royal authority in each county, was often found to be in league with particular noble interests.

If this highlighted the problem, it also suggested a solution. As the disruption caused by the over-mighty subject became chronic, the response of the crown was not to abandon the county system but rather to consolidate it on a broader basis. From the time of Henry VI onward, there were attempts to reduce the powers of the sheriffs and limit their tenure of office. This was the beginning of a process by which authority was gradually transferred to the justices of the peace (JPs), the local gentry who served on the commissions at county quarter sessions. It was slow work. Under Edward IV, the JPs were given jurisdiction in criminal cases, but even this relatively forceful fifteenth-century monarch was unable to free himself of dependence on the magnates to govern and administer broad regions of the kingdom.

Henry VII made the most significant breakthrough towards an alternative balance. The Tudors may not have inaugurated a "new" or "middle class" monarchy as was once suggested.[16] But it is clear that, in effect, they brought about a decisive reduction in the power of the over-mighty subject, and this involved the devolution of political authority downward towards the middle ranks of society. Henry VII was in a sense fortunate that his claim to the throne had not been overwhelmed with noble support, which would have left him with obligations. Nor was he burdened by an illustrious royal family expecting suitable situations. He could therefore take the opportunity to free himself of the feudal habit of rewarding supporters with great tracts of his kingdom. He reversed the relationship. Whereas Edward IV had used crown lands as rewards, Henry kept them for money.[17] The considerable amount of land that came into his hands through attainders was given back only piece by piece, and on such terms that kept the recipient indebted and loyal to the crown. To guarantee good behaviour, he made more systematic use of bonds and recognizances. He showed similar circumspection in creating peers, and none were allowed to attain a position of dominance. Although he did not entirely exclude the nobles from his Privy Council or from the Commission of the Peace, he showed a clear inclination to place more trust in lesser figures. It is now accepted that when choosing his most important advisers and officials, Henry tended more than his predecessors to rely on men who were not of the high nobility.[18]

Much recent writing has emphasized the growing importance of the royal court.[19] But the apparently increasing significance of the court was in part an illusion created by the decline of the over-mighty subject. The rival courts and pretenders that had existed for so long were disappearing from the scene. They were helped on their way by the Tudor determination to change the balance of land power in the localities. "Henry VII and Henry VIII showed a marked reluctance to give control of an area to any one man. They reversed the dominant assumption—never previously questioned—that where a magnate had large lands in a particular area he had a right to rule it".[20] That supposition had been the natural consequence of the feudal concept of land and loyalty, and it was not easy to overturn. The councils that were created to administer outlying areas like the North and the Marches were at first difficult to sustain on a permanent basis. But the policy continued with the same commitment and increasing success as the century wore on.

The most important and consistently successful aspect of Henry's approach to administration lay not, however, in the creation of a royal bureaucracy as such, but in the decisive development of the role of the justices of the peace. The justices still represented local landed power, but the great advantage that they offered to the king was that they related more specifically to the county than had the preceding, more personal agencies of law and private jurisdictions. As the authority of the justices increased, so the county structure began to fulfil its potential as the basis of a uniform system of administration and representation. Henry used the JPs to bring the localities under a more objective structure of legal and administrative control. It has been rightly said that the two main areas of legislation under Henry's rule were his programme of economic reforms and his desire to improve law and order. For the latter purpose, "the powers of the JPs were extended in each successive parliament of his reign".[21] Most interesting are the measures that not only extended their powers but also increased their capacity to supervise and control the activities of other local authorities. Thus in 1485 they were empowered to hold inquests into concealments of other inquests. In 1495, they were given the right to punish sheriffs and bailiffs found guilty of extortion. They were also given the authority to prevent the over-mighty from influencing juries and were empowered to change the composition of a jury if they believed that a sheriff had failed to act impartially. In 1504, they were given the right to certify who were retainers and embracers when an intimidated jury failed to do so. At the same time, their capacity to deal with livery and retaining in general was further increased. All these measures show what Henry regarded as the main threat to law and order and what he believed to be the most effective agency for countering it.

His judgement was sound. There is no doubt that these provisions for good order "commended themselves to the knights of the shire who represented the class from which the justices of the peace were drawn".[22] It was important that the gentry were happy to take on this role. They were also given wide responsibilities for regulating the economic life of their communities. The control of the commissions themselves was not as great a problem as sometimes supposed. True, they were locally based "amateurs" rather than salaried officials, but they had, collectively, a more *public* responsibility for the interests of their localities and counties than had often been the case with the magnates. Furthermore, the commissions were reviewed and renewed annually by the king, and as the significance of the office increased in the eyes of the royal government, so did its prestige in the county. Soon no gentleman could afford the ignominy of being dropped from the bench. Henry also began the practice of directing affairs more closely by appointing leading councillors, trusted clerics and expert lawyers to sit on the commissions. This established links of common interest as well as an effective means of central supervision. It was on this basis that "Henry VII could safely take up the office of JP, adapt it to his needs, and develop it into the mainstay of local government".[23]

The policy was a considerable success.

> The enlisting of gentry communities in the business of self-regulation was a more promising road towards better governance than enlisting magnate support . . . the good understanding between the Tudors and the country gentry, and the Tudor determination not to have men on the

> commissions of the peace who were retained of any cloth but their own, really did go some way towards improving matters.[24]

So the commissions of the peace came to be filled mainly by that same class of gentry who were establishing themselves as the ubiquitous focal points of social and economic power in the villages. Now, as JPs they were providing a counterweight to the traditional, personal power of the magnates, and they were comfortable with the new kind of public authority that this involved. "Though it was locally that the justices of the peace had to apply the rules, matters such as the laws concerning riot, wages and prices, the regulation of guilds, and the abuses of livery and maintenance, were not just local matters . . . they were national matters".[25]

One of the most damaging misjudgements in modern historiography has been the assumption that because the administrative network was amateur and locally based rather than a centrally paid bureaucracy, the force of it was essentially localist.[26] The reality is rather different. The factor that distinguished the system of county administration in England was its kingdom-wide uniformity. As the gentry and yeomanry expanded their powers, they were filling out a balanced and unifying framework of administration and providing the basis for a more integrated form of public authority. The mistake is to assume that unity can only be created by central imposition. In truth, this uniquely coordinated system of administration was as much a national as a local phenomenon. It was, by definition, a basis for associating local needs and preferences through a common structure and meeting them by collective provision. It could work with considerable efficiency as a form of government. The significance of this is recognized, at least in part, by some historians.

> The county officials below the rank of assize judge were, apart from the feudatory, unpaid and for the most part untrained . . . [S]heriffs, JPs, high constables and parish constables were all members of their own communities as well as officers of the crown . . . . Such a system had a strength which centralised bureaucracies lacked, for when it worked as intended the officials provided an essential contact between the crown and the localities.[27]

There is an important truth in this, but it is only half the truth, and the other half was eventually to have a more dramatic impact on the political life of the kingdom; an equally productive but more challenging form of contact was that which developed *between* the localities themselves. A system of balanced local forces, working within an inclusive and integrated network of administration and representative consent, had an obvious potential for developing a coherent national perspective of its own. It will become clear that the eventual problem for the royal government arose not from disruptive localism but from the assertion of an alternative, independent national viewpoint.

For most of the Tudor century, however, these systems worked essentially to the advantage of the crown. They constituted an efficient and stable tool of government, which facilitated the provision of a long series of major national reforms and reformations with remarkably little fuss. The secret lay in the coordination of aims as well as agencies. The process of the break with Rome and the dissolution of the monasteries is a good illustration. One of the most interesting questions about the early stages of the English Reformation is how such a dramatic transformation was carried through with so little apparent difficulty. It is explained in part by the developments in local government that have already been outlined—the encouragement, especially by Henry VII, of the county justices of the peace as the mainstay of administration and law in the localities. By the 1530s, they were settled and successful in that role, and they were ready for Thomas Cromwell to use for other, more dramatic purposes.

But perhaps the most important factor of all was that the secularization of the church lands could be pursued as a kind of national project. The sell-off of the abbey lands to the laity played to the most basic appetites and interests of the gentry—their love of property. This gave them an irresistible stake in Henry VIII's Reformation and guaranteed its success. The broad body of aspiring landowners could be directly involved in the redistribution of the vast properties of the church in their favour. Thus the valuation of the church lands in 1535 was carried out under the auspices of the county commissions. The same people supervised and policed the confiscation of monastic property that began the following year. All over the country, the gentry and yeomanry readily executed the redistributions that would make about a quarter of the land of England available for purchase.

Geoffrey Elton described the relative ease with which these literally world-shattering changes were consequently carried through. Neither Henry nor Cromwell was squeamish about applying extralegal remedies, but they were rarely required. The standard provisions of parliamentary law, implemented through the gentry and the justices, proved perfectly adequate. Cromwell's most distinctive instrument was the very civilized procedure of circular letters to the JPs, quietly and calmly coordinating the practical measures that he thought necessary to sustain a process on which the great majority of the landowning laity was agreed.[28]

The problem for the royal government would come in later reigns, when the policies of the crown were no longer so closely synchronized with the general desires of the political nation and might even be seen as directly contradicting them. In fact, the new network of administration and representation soon showed itself capable of taking a very effective opposition stance. When Mary Tudor sought to restore the church lands to their original ecclesiastical owners or to establish the king of Spain with rights to the English throne, the gentry, through the parliamentary system, just said no.

NOTES

1. P. Anderson, *Lineages of the Absolutist State*, (London 1979), p. 243; for France, see D. Parker, *The Making of French Absolutism*, (London 1983) and *State and Class in Ancien Regime France*, (London 1996).
2. P. Anderson, *Lineages of the Absolutist State*, pp. 140–2.
3. TNA, SP 16/126/44.
4. J.R. Lander, *Government and Community: England 1450–1509*, (London 1980), pp. 42–3.
5. J. Guy, *Tudor England*, (Oxford 1990), p. 352.
6. Ibid.
7. Thomas More, *Utopia*, 1516, ed. J. Warrington, (London 1978), pp. 26, 45.
8. Robert Crowley, "The Way to Wealth", (circa 1550), *The Selected Works of Robert Crowley*, EETS, e. series, XV, (1872), p. 147.
9. Thomas Starkey, *A Dialogue between Reginald Pole and Thomas Lupset*, (1533–1536), ed. K.M. Burton, (London 1948), pp. 89, 143–4, 158.
10. *A Discourse of the Commonweal*, (1549), ed. M. Dewar, (Virginia 1969), pp. 49, 55–9, 93.
11. *Piers Ploughman his Exhortation to the Lords, Knights and Burgesses of the Parliament House*, circa 1550, BL 1653/8, unpaginated but as counted, pp. 2, 4, 12, 15.
12. Sir Thomas Smith, *De Republica Anglorum*, (1562–1565), ed. M. Dewar, (Cambridge 1982), pp. 78–9.
13. Richard Hooker, *Of the Laws of Ecclesiastical Polity*, ed. C. Morris, (London 1968), I, pp. 171, 194.
14. Quoted in J.D. Mackie, *The Earlier Tudors*, (Oxford 1966), p. 27.
15. Ibid., p. 28.
16. Originally suggested by J.R. Green [*A Short History of the English People*, (London 1876)]. It was taken up in the work of A.F. Pollard. One of the leading critics was S.B. Chrimes [*Henry VII*, (London 1972)], suggesting that Henry merely produced a more efficient version of the traditional approach.
17. A. Grant, *Henry VII: the Importance of His Reign in English History*, (London 1985).
18. For instance, C.S.L. Davies, *Peace, Print and Protestantism*, (London 1977), p. 108.
19. S. Gunn, *Early Tudor Government*, (Basingstoke 1995); D. Loades, *Power in Tudor England*, (Basingstoke 1997).
20. S. Gunn, *Early Tudor Government*, pp. 42–4.
21. J.D. Mackie, *The Earlier Tudors*, p. 196.
22. Ibid., p. 201.
23. G.R. Elton, *England Under the Tudors*, (London 1967), p. 60.
24. M. Keen, *English Society in the Later Middle Ages*, (London 1990), p. 212.
25. Ibid., p. 156.
26. The idea that the challenge to the crown was merely localist is summarized in J. Morrill, *The Revolt of the Provinces*, (London 1980), and C. Russell, *Parliaments and English Politics*, (Oxford 1978).
27. P. Williams, *The Later Tudors*, (Oxford 1996), p. 154.
28. G.R. Elton, *Policy and Police*, (Cambridge 1972).

9 The National Expansion of the Middling Sort, and the Relevance of the Rise of the Gentry

The development of the county administrative network was facilitated and structured by the great increase in the number and substance of the gentry as a class. The rise of the gentry is perhaps the single most notable idea in English historiography. It outlined a substantial socio-economic basis for the centrally significant event of the Civil War between king and parliament in the mid-seventeenth century and offered a balanced explanation of the most characteristic aspect of our political past—the emergence of representative institutions. Richard Tawney popularized the phrase "rise of the gentry" as a way of describing the access of strength that enabled a broad landowning class to generate the parliamentary challenge to the crown in the 1640s. He suggested that the gentry were better equipped to take advantage of the Tudor inflation than were the high aristocracy. Lawrence Stone supplemented the thesis by postulating a "crisis of the aristocracy" on a similar basis.

They were not working on a blank sheet. The idea that the ruling elite in society were being challenged by the rise of the groups below had a pedigree dating from the Civil War period itself. In the 1650s, James Harrington bore witness to a process by which "the people" had grown in economic strength at the expense of the crown and the nobility, and he saw this as underpinning the downfall of the royal government in 1641.[1] Christopher Hill cited a list of other leading contemporary commentators who perceived the same occurrence, and he concluded that it was something of an orthodoxy. We can add the testimony of Lucy Hutchinson, who also observed that the nobility had "shrunk into empty names" while "the interest of the people" made a great economic progress, which "cast the balance clear on their side", and they came rolling in upon the crown.[2] Previous chapters in the present volume have described the full extent of that "extraordinary progress" and the strongly expanding socio-economic base of the gentry and yeomanry, of which there is no doubt. To traditional historians, it seemed logical to suppose that contemporaries were also correct about the consequences, and that these developments must in some sense have created the capacity and incentive for MPs to question the policies and powers of the crown in the early seventeenth century, and eventually for the gentry and the middling sort to confront the king militarily in the name of parliament in the 1640s. This perspective gave the most distinctive features of English history the strength of powerful roots.

There was, however, a counter-veiling tendency among historians who felt a personal and political aversion to some of the ways in which history might be given coherent form or thought to be "advancing". In particular, they disliked the idea that the rise of the middling ranks might be taken as substantiating the concept of a bourgeois revolution, as suggested in Marxist theory. So the "revisionist" school of the 1970s and 1980s set out to reject these connections in every way they could. They employed a method of counter-logical reasoning and a highly restrictive and selective focus in order to try to undermine the general force of the traditional view. Hugh Trevor-Roper was first in the field against Tawney, with a kind of mock thesis that the Civil War was induced by the desperation of the declining gentry. A more sophisticated critique came from J.H. Hexter, who inaugurated the standard revisionist technique of narrowing the definition of terms and closing out the real balance of the evidence. Particularly misleading was his attack on the "myth of the middle class in Tudor England". It is not difficult to deny the existence of a discrete, modern-style middle class in the sixteenth century. But this should not be allowed to obscure the related developments that were actually taking place. There clearly *was* a developing bourgeois *context* that formed a crucial stage in the emergence of capitalism, involving the rise of commercial farming, new dimensions of marketing and the assertion of trading freedoms and absolute property rights. To conceal this was to hide an important truth. Similarly, although there was some force in the accusation that Tawney had used the counting of manors in a somewhat clumsy fashion, this weakness should not have been taken to invalidate the basic idea. In fact, when the method was reapplied in a study of landholding in Norfolk, it proved to be powerful, showing that by 1555 two-thirds of the old monastic lands had come into the hands of the gentry, with the remaining third shared between the crown, the nobility and the church, which graphically underlined the justice of Tawney's general observations.[3] A more valid criticism of the Tawney/Stone thesis is that the gentry and the nobility were in many respects from the same landowning class, and there is no reason to suppose that the nobles were inherently less capable of profiting from the general increase in agricultural prices and land values. It is clear nonetheless that the groups below them did significantly better, in both relative and absolute terms. The yeomanry and the lesser gentry had the lowest overheads and the highest profit ratios, and they were the greatest in number. As a class, they were in the best position to make the biggest leap up the socio-economic ladder.

The rise of the gentry and the middling sort has now been affirmed in so many ways that no amount of subterfuge can conceal it. But some historians still seek to assert that it cannot be taken as causative to the English Revolution because most of the greater gentry fought on the side of the king in the

Civil War.[4] This, again, is a simplistic and selective argument, which depends on treating the Civil War in isolation and ignoring other important preceding aspects of the challenge to the crown. Many of the greater gentry may indeed have discovered a residual loyalty to the king when conflict became imminent, and parliament itself sometimes appeared in a more dictatorial light. But as we shall see in detail ahead, in the years since 1600 the gentry had displayed a readiness to undermine the royal prerogative in almost every area, most consistently by their challenge to the king's control of the customs dues. And by the 1620s, they were making royal government impossible in the field of foreign affairs by refusing to fund anything but their own preferred policy. This growing capacity for radical initiatives through the parliamentary forum must be recognized if we are to see the rise of the gentry in its full and true perspective, as shaping the challenge to the crown.

In fact, the radical assertiveness of the gentry found specific expression in the first stages of the programme of the Long Parliament. I have described elsewhere the importance of the Triennial Act of February 1641, which was a core constitutional innovation, and the priority measure of that momentous gathering.[5] Revisionist historians have circumvented the significance of the Triennial Act by restricting their focus to later, less controversial measures. But in truth it was the central proposition of the parliamentarian movement and the true beginning of the English Revolution. Furthermore, it had the virtually unanimous and committed backing of the gentry in the Commons. Its effect was to remove the royal prerogative over the summoning of parliament and enforce regular and automatic meetings. This reflected parliament's long pursued ambitions to enhance the force of representative rights at the expense of the prerogative, across a range of governmental fields. The act established the representative assembly and representative concept at the centre of political life, and this specifically related the rise of the gentry to the English Revolution.

So indeed did the measure that fulfilled parliament's most specific and long-standing radical ambition—the elimination of the royal prerogative over the customs dues. This was achieved by the Tonnage and Poundage Act of May 1641, which deprived the king of any substantial income independent of parliament. This too was a priority constitutional innovation that had the unanimous backing of the gentlemen MPs in parliament. It has not received much attention from modern historians, but Samuel Gardiner recognized its force: "By this bill, Charles surrendered for ever his claim to levy customs dues of any kind without a parliamentary grant".[6] This and the Triennial Act were the particular measures that gave substance to Harrington's assertion that "the dissolution of this Government caused the War, not the War the dissolution of this Government".[7] The king was obliged to fight the Civil War to try to recover the authority destroyed by these first, unanimous radical moves of the Long Parliament.

The conjunction of these two early revolutionary acts highlights the connection between economic factors and the desire to enhance the position of parliament. There were various distinguishable motives for wanting to establish the representative assembly with a permanent place at the centre of political life, the intention that was embodied in the Triennial Act. One purpose was to ensure the systematic exercise of parliamentary consent in all matters of public finance and money raising, like the customs dues. Another was to guarantee the regular provision of parliament's legislative service, which MPs and their constituents now saw as vital for the efficient regulation of political and socio-economic affairs, and essential to the good government of the country. In 1641, the parliamentarian leader John Pym would hail the Triennial Act as ensuring the continual exercise of parliament's power of legislation, which had become in his view, "that which makes and constitutes a kingdom". Sovereign, representative law was crucial for their practical needs and seminal to the self-awareness of the political nation.

The application of a single, sovereign representative law to the most important public priorities was the vehicle through which the political nation gathered and pursued its self-interests. This provides a general, collected definition of the motives behind the demand for automatic parliaments and relates it to our central theme of the changing conception of "the land". An important and neglected cause of the English Revolution was the desire of the political nation simply to establish more efficient ways of giving effect to the independent perception of the public interest that had developed over preceding decades. This was the coordinated force of the specific public priorities that were being identified. It was reflected in the way that the gentry had come to embody the preferences of the kingdom at large and in the extent to which the gentlemen MPs in parliament were now willing and able to identify a unitary national interest, independently of the policies of the crown.

In all these ways, the gentry were at the heart of the political dynamics and desires of the political nation, and their role depended upon them becoming a national class. As already suggested, this had a visible, physical aspect on the ground, as the gentry proliferated, and spread out through the counties and the county networks and the villages. We can visualize more clearly how they were filling out the land politically if we recall that this was a kingdom with no substantial towns except the capital, but just a great number of villages, each now coming to be dominated by its squire. Their advance was not just a rebalancing of society and local status. This was secular power completing the occupation of the land and fixing its own character upon it. The remarkable revolving network of fluid and open associations that had centred on the universal church was draining away, as social and economic power was funnelled into the hands of the gentry. In their direction had come the bulk of the church lands, including the tithes and the control of church livings. This was indicative of a change in the character of authority. Suddenly the dominant presence and defining purpose of the church had disappeared. With almost equal suddenness, all the communities in England had acquired a prominent focus of secular power. The nation state was taking solid form on the ground.

In this instance, the statistics tell the story. The situation in Shropshire, where there were forty-eight gentry families in 1433 and four hundred and seventy in 1623, reveals not only the scale but also the force of the transition. In the medieval period, the class did not really possess the substance to provide a vehicle of administration. In fact, at that time the county gentry scarcely existed as a consolidated body. Their position was still defined essentially as a scattered order of knights and bachelors who held land in military service to an overlord. But by the end of the sixteenth century, they were often something like ten times as numerous, and they were physically filling out each county, in the same way that they were coming to be the dominant presence in every village. The massing of the gentry was repeated all over the kingdom. There were at least four hundred gentry families in Norfolk by the 1580s, and no less than a thousand in Kent by 1600. J.T. Cliffe found six hundred and forty-one in Yorkshire in 1603 just on the basis of the herald's visitation and would have registered half as many again if he had taken the less strict measure of those who simply went by the title of gentry. The crucial further implication is that as the gentry spread through the counties as landed powers, they were also filling out "the land". And as they came to counterbalance the high nobility, both in the countryside and in the royal councils, it was not just a process of devolving power to a lower social strata; it was placing political authority on a more public basis. The commission of the peace was the forum in which the force of the gentry was coordinated at county level, and the role of the counties as equal units in a uniform structure of national administration and representation meant that the rise of the gentry would also find national-political expression.

The gentry were not simply commandeering the vast acreage and predominant cultural position formerly held by the monastic houses; they were also occupying the interstices in the less consolidated patterns of power that had been exercised by the feudal nobility. The power relations of the feudal era had been more personal than public. They were fluid and open, like other aspects of the medieval world. But the increasing density of the gentry on the ground was closing down that space. They now had a powerful presence in every community in the land. This gave them a universal face, as the instruments of a more complete public authority, both within their counties and collectively as the county representatives of the kingdom. Thus, the redistribution of landed strength into the hands of a much broader class of gentry was the foundation of a new balance of power, which was ultimately more national in character.

An important implication in terms of the distribution of political influence was clearly set out by Lawrence Stone in the most persuasive section of his book, dealing with the devolution of power. One sense in which it is certainly fair to speak of a decline of the aristocracy is that there was a substantial reduction in the numbers and powers of the greater territorial magnates. Stone found that the numbers of noble families holding seventy or more manors fell from eighteen out of sixty-two in 1558 to six out of one hundred and twenty-one in 1641. These figures are clear enough to be significant, and he was right to say that "the top level of the English social pyramid had been substantially reduced between the accession of Elizabeth and the outbreak of the Civil War".[8] This conclusion has been confirmed more recently in Mark Overton's summary. Between 1436 and 1690 the share of land held by the greater nobility remained static between 15% and 20%, while the share held by the gentry doubled from 25% to 50%. By the later seventeenth century, the gentry and the yeomanry together held three-quarters of the land of England.[9]

The transfer of landed strength had further consequences for the balance of office holding. The increased weight and ubiquity of the gentry on the ground made them available to the crown as an alternative administrative option, significantly different in character from the feudal relationships on which the monarchy had previously relied. The prominent architectural presence of the first stylized manor houses and the rebuilt farmhouses reflected a new framework of authority, an extension of power points much greater in number but without the personal, regional dominance of the great barons. This was the basis of a political and administrative pattern more structurally integrated and publicly coherent than anything that had existed in the medieval world. The monasteries had referred themselves to the universal church. The castles had reflected the private military might of the magnates. Both these pillars of the medieval scene had, by definition, militated against the consolidation of a unitary nation state. But they were now gone, and, by contrast, the rise of the gentry and yeomanry tended to reinforce the potential for uniform administrative and representative structures, both within the county and in the kingdom at large.

Thus the Tudors managed to free themselves of the problem of the over-mighty subject to a significantly greater extent than was achieved by their predecessors. They did it principally by placing more trust and authority in the uniform county structures represented by the gentry and the JPs. The process had begun in earnest in the reign of Henry VII and was completed in the reign of his granddaughter Elizabeth, who made sure to create no over-mighty subjects and was not backward in disposing of the few remaining contenders for that status. Yet she seemed to lack either the desire or the ability to stem the torrent of claims for heraldic arms from the gentry, which created the rapidly expanding pool from which a new administrative class could be drawn. So when the Duke of Norfolk fell from grace in 1572, he was not succeeded like for like by a figure of comparable magnate influence. His traditional position of personal authority in the region was not to be replicated or revived. The vacuum was filled by a larger number of gentlemen of lesser rank, who took control of affairs through the levers of county administration.[10] The ancient feudal power of the Percy and Neville families in the north suffered the same fate. The time-honoured practice of politics by baronial rebellion had come to an end. The gentry were not backward in filling the void. "Where there were many new gentry, like Suffolk and Northamptonshire, a substantial group of gentlemen felt confident for the first time to challenge the traditional leadership of the great families".[11] Thus, in Elizabeth's reign "there was a decisive shift to dependence on the squirearchy and gentry". And by the end of the century, "the diminution of magnate power had produced a wider diffusion of authority among provincial landowners".[12]

But perhaps the most interesting and instructive illustration of the way that the occupation of the physical land was filling out into the unified, political structure of a national land was the incorporation of the yeomanry into the networks of power. This was the class that had benefited most of all from the seller's market of the sixteenth century. Previous chapters have identified the "chief farmers", who had a minimum of sixty acres and a consistent marketable surplus and who were best placed to exploit the economic conditions of the time. Many yeomen had risen socially to swell the ranks of the gentry. Others had flourished on their profits as tenant farmers. Their affluence was conspicuous in the Great Rebuilding, which turned humble farmhouses into impressive adornments of the countryside. Robert Allen has hailed the yeomen farmers as the principal instruments of the commercial and technical development that distinguished the performance of English agriculture in the early modern period.[13] Many had managed to add significantly to their holdings and to their status in the community. These were the groups that Lucy Hutchinson had seen as making "an extraordinary progress" and that James Harrington referred to as "the people", the small gentry and yeomanry whose success had changed the balance of landed wealth.[14]

Jane Whittle has described the context of independent activity through which the yeomen rose: "The significance of the period 1440 to 1580 . . . lies in the freedom, prosperity and . . . lack of landlordly interference experienced by the rural population, and in the fact that an economy generated by small landowners unburdened with heavy exactions by state or landlord could promote the development of capitalism".[15] On this basis, they acquired a stake and an accepted position not only in the affairs of their neighbourhoods but also in the social and political structures of the kingdom. They were the churchwardens and constables, and, equally important, they were often the voters and subsidy men too. "In many villages they constituted a local elite, drawing closer in economic and political interest to the gentry above them".[16] Their part was specifically recognized at the time. Sir Thomas Smith, in his authoritative analysis of the structures of English government at the beginning of Elizabeth's reign, gave the yeomanry a very specific and significant place among the parties by which England was "governed, administered and maintained". There were "three sorts of persons" charged with this responsibility. One was the prince. The second was the gentry (under which heading, interestingly, he included the nobility). "The third is named the yeomanry: and each of these hath his part and administration in judgements, correction of defaults, in election of offices, in appointing tributes and subsidies, and the making of laws".[17]

A century earlier, an analysis of government only needed to mention the prince and the nobility.[18] This reflected an essentially personal and feudal power relationship. The governing "order" was the king and the great magnates who were his lieutenants. The knights comprised the core of the subordinate military "order". The agricultural "order" was composed of the husbandmen, who had no connection at all with the structures of power. But by the reign of Elizabeth, the medieval concept of determining responsibilities on the basis of the distinct "orders" of society had been displaced to the point of disappearance. Smith could therefore conceive of government as an interlinked, kingdom-wide network of administration spreading out through the gentry and the yeomanry, with no separate role for the nobility. This was the basis for a self-conscious public unity and for the open, physical land coming to be seen as a national political land.

Care should be taken not to completely erode the importance of the class division between the gentry and the commonalty. It remained the clearest line of social distinction and socio-economic demarcation. It was the contrast between on the one hand those who worked and on the other those who lived off their rents and whose function was defined increasingly in terms of administrative power. Yet for practical purposes, the line seems to have been very fluid. For instance, one-third of the JPs in Elizabethan Essex were drawn from non-gentry families. In the world of administration and indeed of commercial activity, the gentry and the middling sort were often to be found working together.

It has been suggested by Steve Hindle that a feature in state formation was the widespread use of the legal system by the "middling sort", principally as a means of maintaining order in their localities.[19] The increasing resort to law in the sixteenth century is well known, and there is no doubt that the middling sort were participating in this as in other things. But it should not be supposed that the main interest of the farmers and merchants in the processes of law was for the defensive and inward-looking business of keeping order. There were stronger, general factors that served to consolidate the perception of a "state". We need to recognize the constructive and aspirational forces that were the real impetus for change. It was not just the maintenance of order that brought the middling sort into closer association with the gentry. Of greater import was their common interest in the proactive lawmaking function. The flood of legislative initiatives that flowed forth from the constituencies in the reign of Elizabeth were mainly concerned with advancing the economic interests of the yeomen and the townsmen. The inducement was their understanding of the definitive sovereign provisions that they hoped would meet their needs. Thus, a more definite and positive genesis of the idea of a nation state is to be found in an awareness of the unifying systems and concepts of representation through which the law was *made* and the benefits that it was thought to offer. Smith's inclusion of the yeomanry among those involved in "the making of laws" is apposite. This, more than anything else, brought them into the same administrative world as the gentry. The real framework for the development of the state was the central participation of the gentry and the middling sort in the structures of sovereign legislative power in order to serve the aims and ambitions of the "entire society". And as will become clear ahead, the national force of these relationships was underlined by the fact that parliamentary statutes were made specifically by the authority of that same "entire society".

NOTES

1. R.H. Tawney, "The Rise of the Gentry 1558–1640", *Economic History Review 11*, (1941), pp. 1–38; L. Stone, *The Crisis of the Aristocracy 1558–1541*, (Oxford 1965); C. Hill, "James Harrington and the People", in *Puritanism and Revolution*, (London 1968), p. 289.
2. L. Hutchinson, *Memoirs of the Life of Colonel Hutchinson*, ed. J. Hutchinson, (London 1846), p. 75.
3. H. Trevor-Roper, *The Gentry 1540–1640*, Supplement to *Economic History Review*, I, (1953); J.H. Hexter, *Reappraisals in History*, (Aberdeen 1961), pp. 71–162; T.H. Swales, "The Redistribution of Monastic Lands in Norfolk", *Norfolk Archaeology XXXIV*, (1966), pt. I, pp. 14–44.
4. B. Coward, *The Stuart Age*, (London 1994), p. 151.
5. G. Yerby, *People and Parliament*, (Basingstoke 2008), chap. 6, "The Triennial Act of 1641".
6. S.R. Gardiner, *History of England*, (London 1884), IX, p. 400.
7. J. Harrington, *Oceana and other works of James Harrington*, ed. J. Toland, (Dublin 1737), p. 70.
8. L. Stone, *The Crisis of the Aristocracy*, pp. 153, 264.
9. M. Overton, *The Agricultural Revolution*, (Cambridge 1996), p. 167.
10. A. Hassell-Smith, *County and Court: Government and Politics in Norfolk 1558–1603*, (Oxford 1974).
11. C. Palliser, *The Age of Elizabeth*, (London 1983), p. 89.
12. L. Stone, *The Crisis of the Aristocracy*, p. 256; P. Williams, "The Tudors and England", *History Today*, September 1985, p. 31.
13. R. Allen, *Enclosure and the Yeoman*, (Oxford 1992), p. 14.
14. Lucy Hutchinson, *The Life of Lucy Hutchinson*, p. 75; and, C. Hill, "James Harrington and the People", in *Puritanism and Revolution*, (London 1958).
15. J. Whittle, *The Development of Agrarian Capitalism*, (Oxford 2000), p. 315.
16. N. Carlin, *The Causes of the English Civil War*, (Oxford 1999), p. 120.
17. Sir Thomas Smith, *De Republica Anglorum*, ed. M. Dewar, (Cambridge 1982), p. 77.
18. J. Lander, *Government and Community: England 1450–1509*, (London 1983), p. 42.
19. S. Hindle, *The State and Social Change in Early Modern England*, (Basingstoke 2000).

10 "The Authority of the Whole Realm"

Parliamentary Law as the First Principle of Representative Rights and National Sovereignty

The English parliament possessed a distinctive capacity as the vehicle for collective interests, and this gave the rising classes of yeomanry and gentry a very effective national forum. The uniform network of county administration had a public dimension, not only in its inherently balanced kingdom-wide structure, but also because it was the basis of a similarly coordinated system of elections to a representative body with a uniquely national perspective. This assembly boasted a particularly significant unifying principle, for it was parliament's ability to reflect the consent of the "entire society" that was accepted as the authority for the making of sovereign law. The broad popular reception of this concept gave the idea of representative rights a prescriptive philosophical force and created a platform for their further extension. A right of consent could be asserted to challenge the prerogative in various fields, such as the raising of customs dues. This was a particular public priority that contributed to the growing need of MPs and their constituents to give voice to the emerging, independently conceived national interest in general.

So the political nation evolved and revolved in large part around the institution of a parliament that was unique in Europe in its capability to represent the interests and sentiments of the kingdom as a whole. The representative system worked mainly through the rising classes identified in previous chapters. The newly affluent yeomen and successful farmers constituted the bulk of the county electorate. Furthermore, by the early seventeenth century, that electorate was much more substantial than would normally be expected in a pre-democratic age. The great Tudor Inflation had made the traditional 40 shilling freeholder qualification virtually meaningless, just as it had breached many other age-old limits. And during the early decades of the seventeenth century, the House of Commons consistently opted for the broadest definition of the urban franchise when disputes arose. It has been calculated that in consequence, by 1641, those entitled to vote, "may have formed between 27 per cent and 40 per cent of the adult male population".[1] The electorate had a more genuinely national substance than might have been supposed.

Those who were chosen as MPs were now almost invariably the landed gentry. This was another important factor in consolidating the national scope of the assembly. The special character of the English parliament derived at root not only from the uniform structure of county representation but also from the fact that by the sixteenth century, the county gentry had come to virtually monopolize the composition of the House of Commons. Representative assemblies in other kingdoms still continued to reflect society in typically medieval fashion as a series of separate "orders"—the ecclesiastical, the military and the mercantile. Under this traditional arrangement, the merchants comprised the Commons House. In England, however, the Commons had for some while included the gentry who represented the counties. In the later fifteenth century, the status of the Commons was enhanced as the House acquired an executive part in the legislative process, joining the king and lords as assenters to the measures rather than as just petitioners. The gentry were also assuming responsibility for implementing the statutes, so they sought to maximize their position in the lawmaking body. Their takeover of the borough seats was almost complete by the late sixteenth century. Under the traditional criteria, there would have been four burgesses to every country gentleman, but this position was now reversed and there were four gentry to every townsman. The gentry were also engineering a substantial increase in the size of the assembly in order to give themselves yet more scope. At the beginning of the sixteenth century, there were two hundred and ninety-six seats, and by the end there were four hundred and sixty-two. Much of this inflation came in the reign of Elizabeth, when there was "a rapidly expanding desire to sit in parliament".[2] It was "from the country gentry that the demand came".[3] It is indicative of the force of the development that the usually cautious Queen was either unable or unwilling to resist it.

The House of Commons thereby lost its original character of a medieval mercantile chamber and became predominantly composed of gentry. So the same local landowners who were consolidating their power base in the villages and the counties and who had been adopted the instruments of public administration as JPs had also extended the scope of their role as MPs, and had come to monopolize the greatly increased membership of the Commons. This had advantages and disadvantages for the market towns that lost the direct use of their franchise. Their representatives were no longer their own townsmen with an intimate grasp of their concerns but were instead more powerful figures with a greater influence in the national arena as it developed. This was on balance beneficial to the towns because the gentry and the mercantile classes were of one mind over such crucial aims as bringing the customs dues under representative control.[4] This exemplified indeed the kind of common, public priorities that were emerging.

The gentry's takeover turned the Commons into a body with a unique ability to reflect the collective opinions and unitary interests of the kingdom. Their central responsibilities as JPs and MPs were mutually reinforcing. They were elected on the strength of their standing in the county, which was in turn enhanced by their role as national representatives. The authority of their assembly was itself increased by its place as the defining element in a new level of sovereign lawmaking power, based on its ability to reflect the "entire society". Accordingly, from the mid-sixteenth century, statute was employed routinely to reform the most vital aspects of the life of the kingdom. And having played a central role in making the laws, the gentry returned to their counties to monitor and enforce them. This was the core dynamic of a new kind of public authority.

The developments in the nature of legislative power that enhanced the status of parliament and representative rights also created a political and psychological basis for the idea of national sovereignty. The break with Rome and the removal of the independent, universal church from within the realm were by definition crucial steps in the completion of a unified, sovereign secular state. Just as the state was given physical consolidation by the lay appropriation of the church lands, so the ejection of the authority of Rome from the kingdom's shores was accompanied by a formal declaration of complete integrity and independence. Like the undisguised lust for the church lands, the concept of secular independence felt no need to hide itself or to camouflage its intent. It was proclaimed in the preamble to the Act in Restraint of Appeals, which confidently asserted that "this realm of England is an empire". This meant that the kingdom was now completely autonomous, governed by one supreme head and king, to whom the whole body politic, temporal and spiritual, owed obedience and who possessed plenary power to "yield justice" to his subjects without restraint of any foreign prince or potentate.[5] The clarity of the concept is indicative of the coherent principle behind it. This was not simply the gathering of regnum and sacerdotium under the same umbrella: The kingdom was being presented "as an organic whole".[6] In fact, it was being conceived, for the first time, as a nation state, and the cement was the unifying dynamic of parliamentary statute. All law, civil and ecclesiastical, should be made in parliament, said Christopher St German, because parliament represented "the estate of all the people within this realm, that is to say of the whole Catholic Church thereof".[7]

So the break with Rome had another vital contribution to make to the dynamics of the consolidated nation-state. This was the creation of a new level of overriding, omni-competent statute law. The Act in Restraint of Appeals declared itself to have been carried through "by his royal assent and by the assent of the Lords spiritual and temporal and by the Commons in this present Parliament assembled, and by authority of the same".[8] Thus the measure had been implemented specifically through the kingdom-wide force of parliamentary statute, which in the process established an indisputable ascendancy over all other types of law—such as canon and customary law—that had previously possessed independent status. Parliamentary law had acquired an indivisible supremacy, which was as significant as that assumed by the king.

The sovereignty of parliamentary law was the binding judicial strength of the newly independent nation. But it was more than just an innovative legal technique. The process worked specifically to enhance the concept of national representation as the basis of the integrity of the realm. Consent became the defining force of sovereign legislative power. Sir John Fortescue had noted in the mid-fifteenth century that in England laws and taxes required consent. But he had not placed the same stress on the centrality of the representative concept as it now came to assume. The idea that law was binding upon all because it was made by the consent of all became current among the legal fraternity at the end of the fifteenth century, when the House of Commons began to be routinely included in the wording of statutes as having "assented" to the act, along with the king and the Lords. The concept thus established was available to Thomas Cromwell to use as a means of superseding the laws and independent authority of the church. But it was only when that revolution had been achieved and the representation of "everyman" had become the authorizing principle of the one, indisputable sovereign law that the idea could be taken broadly into public consciousness. The principal architects of the transition, Cromwell and the lawyer Christopher St. German, shared a vision of a consolidated secular state, the unitary force of which derived at root from the capacity of parliament and parliamentary law to reflect the will of the entire kingdom. This principle thereby served to embody, as well as to create the new nation.

So the concept of the omni-competent sovereignty of parliamentary law was as much the instrument as the result of Henry's break with Rome. Only parliament could offer a settlement that was, by definition, accepted by everybody in the kingdom. Only parliament could deliver the "universal" moral and judicial authority of the realm that Henry needed as an alternative to the universal authority of Rome. The moral force of the representative concept was elucidated by St German in the mid-1520s, when he began to set out the case upon which statute law should supersede the law of the church. "For it cannot be thought that a statute that is made by the authority of the whole realm, as well as of the King and the Lords Spiritual and Temporal, as well as of all the Commons will state a thing against the truth".[9] The capacity to embody a national "truth" other than the truth of the church was the revolutionary support that parliament could provide. And this was the basis on which the representative capacity of the assembly gave birth to a groundbreaking concept of fully sovereign law.

In the "New Additions" to his work in 1530, St German spelled out the implications as a concept of the nation state. "The king in his parliament is the high sovereign over the people, which hath not only charge of the bodies but also the souls of his subjects".[10] John Guy concluded that "this virtually amounted to a full blown theory of parliamentary sovereignty".[11] If we are thinking of sovereignty as the function of an omni-competent legislative power, this was certainly the case. And it was the core dynamic of the newly unified kingdom. In 1535, St German reiterated the rationale. All law, civil and ecclesiastical, should be made by king-in-parliament, "for the parliament so gathered representeth the estate of all the people in this realm".[12] In these words we can hear the distinct "states" or orders of medieval society being overlaid, as a new defining identity—the place of everyone in the single state of the realm—took shape.

Unlike Cromwell and St German, Henry himself was not entirely comfortable with the idea that his supremacy had been made in parliament, and he clung to the notion that the assembly had merely declared the divine truth of it, in time-honoured medieval fashion. But, in fact, once it had been established that only the consent of everyman could bind everyone to law and that therefore only parliamentary law could be sovereign, it was a difficult thing to get round. This became apparent in the reign of Henry's daughter Mary. In her view, the law of the church retained its independent sovereignty, and the parliamentary statutes that had denied it were invalid. But when she sought to divest herself of the supremacy and return it to the Pope, she found that this too could only be done by parliamentary statute.[13] The principle was now settled—sovereign law could only be made in parliament, and only unmade in parliament.

So the nation state of England was born. It was brought into the world by the concept of an unchallengeable lawmaking power that derived its authority from the capacity of parliament to reflect the consent of the entire kingdom. The collective context of sovereign representative law was given expression in the authoritative definition provided in 1564 by the leading expert in the principles and practice of Tudor government, Sir Thomas Smith.

> The most high and absolute power of the realm of England is in the parliament . . . the barony for the nobility and lords; the knights, esquires, gentlemen and commons for the lower part of the commonwealth . . . consult and show what is good and necessary for the commonwealth, and upon mature deliberation every bill or law . . . that is the prince's and the whole realm's deed: whereupon justly no man can complain, but must accommodate himself to it and find it good . . . For every Englishman is intended to be there present, either in person or by procuration . . . from the Prince to the lowest person of England.[14]

Historians have sometimes pointed out that this was not intended to draw a battle line between crown and parliament for the title of sovereignty. But that should not be allowed to obscure the radical potential contained in the concept. The inclusion of "the lowest person in England" was sign and symbol of the fact that the representation of "everyone" had become the defining basis of a process of sovereign lawmaking. True, the legislative body was still king-in-parliament, but by that very token this meant that the monarch was not regarded as possessing the authority to make sovereign law in his own person. And therein lay a significant underlying threat to the powers of the crown, for the notion of an "absolute" power that could only be exercised through the representative assembly was a basis of principle on which the royal prerogative could be challenged in other areas and the demand for rights of parliamentary consent extended, if the need should arise.

Of particular interest in the context of this study is the way that Smith captures the collected public character that lawmaking had acquired. The whole kingdom gathered together to consult on the nature of its aims and problems and on how they could be dealt with by the provision of national sovereign law. It was the "whole realm's" deed and was therefore binding upon all because all had consented to it. Richard Hooker, three decades later, found the phrase that summed up in clear political theory the autonomous national perspective that this involved. "The lawful power of making laws to command whole societies of men belongs so properly unto the same entire societies".[15] Hooker also underlined the radical thrust of the proposition that the monarch could not make sovereign law of his own authority. "For any prince or potentate of any kind whatsoever on earth to exercise the same of himself . . . is no better than mere tyranny . . . [L]aws they are not therefore which public approbation hath not made so".[16] So by these authoritative views, in England by the late sixteenth century, only the complete, represented community could make sovereign law. Thus, the overriding sovereignty of parliamentary law, as made by the "entire society", became the most important conceptual framework for the emergence of the idea of a national land. This above all was what defined the kingdom within its own political structure. This was what established the perception that the nation constituted a political force in itself.

Theory was reflected in practice. It is clear that the idea of the absolute sovereignty of parliamentary law was assimilated by the political nation at large. An awareness of the principle was indeed entailed in the nature of the process. Just as people knew that they were bound to pay parliamentary tax by their own consent, so they knew that they were bound to obey parliamentary law because they had assented to its making. They entrusted their MPs with that consent and hoped in return to receive the benefit of definitive national law to serve their interests. The popular reception of the concept was shown partly in the ever increasing public demand for parliament's legislative service, despite the fact that from the late sixteenth century the royal government showed less and less inclination to harness the process on its own account. The special public appeal of sovereign representative law lay in its dual force: it could directly reflect the wishes of the constituencies, locally and generally, and it could meet their needs with uniquely powerful national provisions. So by the end of Elizabeth's reign, the demand for legislation had risen to a level much too great for parliament to satisfy, and few of the initiatives came from the government—they came overwhelmingly from the general population.[17]

The principle was understood not only as a working concept but also at a constitutional level. It is generally acknowledged that the idea of the sovereignty of representative law had acquired wide currency, bordering on unanimity, among the political nation by the beginning of the seventeenth century. Some historians have taken this to suggest that it was "uncontroversial".[18] But, as we have seen, it carried the very radical implication that the monarch could not make sovereign law of his own accord. And although it was a matter agreed upon among the gentry and other politically active classes, it was not uncontroversial to the king, or other royal apologists. On the contrary, they took a different view, with good reason. The establishment of a broad public opinion that sovereign law could only be made in parliament posed an underlying challenge to the status of the crown. The threat was apparent to James I and to the Divine Right theorists among the lawyers and churchmen who followed his line. They were aware that the concept of the exclusive sovereignty of representative law would have to be deflected if the sovereign authority of monarchy was to be kept intact. It is sometimes doubted whether they reflected Divine Right theory proper, which was not fully articulated until later in the century. But the two phases were consistent in at least one crucial respect—the desire to assert that representative consent was not intrinsic or essential to the promulgation of sovereign law. Instead, Divine Right theory held that although the king might, as a concession of the royal grace, admit parliament to a place in the lawmaking process, this was not an inherited right, and the king could still make sovereign law by his own authority if he so desired. This was the position asserted by James himself in his political writings.[19] It was also espoused by civil lawyers like John Cowell. In his book *The Interpreter*, which so incensed opinion in the Commons, Cowell declared that "the prince of his absolute power might make laws of himself".[20] This contradicted the parliamentary principle that sovereign law could only be made with the consent of the representative body. James also found support from the church, whose Canons tended to reassert the view that the monarch was the essential source of law and could legislate alone. So although the sole sovereignty of parliamentary law was broadly accepted by the political nation, it was not orthodox among the Stuart establishment. At the end of the sixteenth century, Richard Hooker's liberal view of the sovereignty of the entire society fell out of favour, and "the doctrine that kings derive their powers from God alone was the orthodox teaching of the early Stuart clergy".[21]

These conservative reactions troubled the parliamentarians more than is usually recognized. A little noted aspect of the Commons *Apology* of 1604 is its concern to rebut the suggestion that the king could make sovereign law by his own authority, outside parliament. Alarm bells had been set ringing by a gloss on the Canons of 1604, suggesting that the king could personally confirm ecclesiastical regulations and give them the force of law. The *Apology* warned that the king "should be misinformed if any man should deliver that the kings of England have an absolute power in themselves either to alter religion . . . or to make laws concerning the same, otherwise than as in temporal causes, by consent of parliament".[22] It was a point on which the gentlemen of the Commons were at once extremely sensitive and extremely firm.

The tension resurfaced in1606, when Sir Edwin Sandys defended the integrity of the system of statute law in face of the proposal of the union of the two kingdoms, which would have attached the different system of law that obtained in Scotland. The Commons' reluctance is usually attributed to a jingoistic objection to all things Scottish. No doubt an element of this was present. But more significantly, there was a basic objection of principle. MPs were anxious to defend the belief that sovereign law could only be made by the "entire society". In Scotland, the concept of the prerogative-based Civil Law was still powerful. Sandys therefore proposed a "perfect union" by which Scottish law would be subordinated to the English parliament. James did not see the need for this. He told them, with his usual lack of false modesty, that they already had a perfect union—himself—and any anomalies could be resolved by his own authority, since in the final analysis, "rex est lex".[23] This encapsulated a basic difference of opinion about the nature of law. James believed that the monarch could still make sovereign law in his own person if he wished. The Commons on the other hand now believed, finally and definitively, that it could only be made in parliament. The perfect union was designed to sustain that principle. If James had accepted this, he would have been undermining his ability to assert his own residual legislative sovereignty.

The reasons that James might not have wished to give full and free scope to the parliamentary position were underlined by Sandys's passionate advocacy of the representative concept. There must be just one parliament for a united kingdom, he said, so that "the whole do join in making laws to govern the whole; for it is fit and just that every man do join in making that which shall bind and govern him, and because every man cannot be personally present, therefore a Representative body is made to perform that service".[24] He spoke of "the great trust committed to the House of Commons by the whole commonwealth".[25] His words elucidated the core political and philosophical dynamic of the independent, public self-consciousness that was now emerging.

As I have written elsewhere, the idea that the "most absolute power" of lawmaking drew its authority from parliamentary consent created a political paradox. It was a tension that held great potential for the extension of representative rights. If that most "absolute" power could only be exercised in parliament, then should not the same apply to other sovereign powers? The principle was to be employed in precisely this fashion in 1610, when leading parliamentary speakers cited it in justification of their desire to remove the crown's right to set customs dues by prerogative. A more complete resolution of the paradox would emerge in the early revolutionary measures of the Long Parliament, especially the Triennial Act, ending the monarch's prerogative over the calling of the representative assembly and defending the move on the same basis of the high significance of the sovereign power of lawmaking, which could "only be used in parliaments" and was that "which makes and constitutes a kingdom".[26]

The radical force of the concept found most clear and interesting reflection in the contrasting views of the French lawyer and philosopher Jean Bodin, who was writing in the 1570s and who inadvertently "predicted" the manner in which the paradox would be resolved. Bodin's purpose was to suggest an antidote to the disorders in which the French kingdom was embroiled at the time. He therefore set out what he regarded as the essential basis of political order. What he produced is accepted by most as the first full-fledged theory of sovereignty. His premise was that sovereign authority should be indivisible. And the principal aspect of indivisible sovereignty was the power to legislate for all without exception or question. Bodin conceived legislative power as the principal sign of sovereignty precisely on the basis that it should be exercised without the need to seek representative consent. "It is clear that the principal mark of sovereign majesty and absolute power is the right to impose laws generally on all subjects regardless of their consent".[27] "If the prince can only make laws with . . . consent . . . it is not he who is the sovereign".[28]

It was a reasonable proposition, and Bodin appears to have been unaware that the situation in England defied this logic and that the sovereign lawmaking power was quite specifically shared between king and parliament. He knew that English MPs were not without some distinctively assertive habits and that, for instance, they were unusually inclined to assume a license to air their views on matters of state. But Bodin still believed that as with other European assemblies, the English Commons were merely petitioners for the grace of the king when it came to the making of laws. He thought that they had a purely non-executive role in the legislative process and that this was illustrated by the fact that they depended on the king to call them into being. The latter assumption was quite correct, but it did not reflect the former as Bodin supposed. In fact, the representative Commons now possessed an essential and defining role in the making of sovereign law, so that in England that power could only be exercised through their consent. By a kind of reverse light, Bodin had illuminated the political contradiction contained in the English concept. He believed that sovereign legislative power must go together with an established and independent position in the political structure. But he knew that the English parliament could not assemble without express royal command, and he thought that this was "sufficient proof" that it possessed no executive part in the making of law.[29] It was certainly reasonable to suppose that a sovereign lawmaking authority should possess an automatic place in the structures of government. And this was the manner in which the English paradox would find resolution. Early in 1641, with the Triennial Act, parliament provided itself with just such an automatic place in the polity, as the necessary accompaniment to its central, executive role in the vital function of sovereign lawmaking.

NOTES

1. D. Hirst, *The Representative of the People?* (Cambridge 1975), p. 104.
2. J.E. Neale, *The Elizabethan House of Commons*, (London 1950), p. 148.
3. Ibid., p. 146.
4. G. Yerby, *People and Parliament*, (Basingstoke 2008), chap. 3, and see ahead, pp. 165–7.

5. *Statutes of the Realm*, 24 Henry VIII, cap. 12.
6. J.D. Mackie, *The Earlier Tudors*, (Oxford 1966), p. 357.
7. Christopher St German, *Answer to a Letter*, (London 1535), p. 102.
8. *Statutes of the Realm*, 24 Henry VIII, cap. 12.
9. Christopher St. German, *Doctor and Student*, (London 1523), p. 300.
10. Ibid., p. 327.
11. J. Guy, *Tudor England*, (Oxford 1988), p. 127.
12. Christopher St German, *Answer to a Letter*, p. 102
13. J. Guy, *Tudor England*, pp. 133–4.
14. Sir Thomas Smith, *De Republica Anglorum*, ed. M. Dewar, (Cambridge 1982), p. 78.
15. Richard Hooker, *Of the Laws of Ecclesiastical Polity*, ed. C. Morris, (London 1969), vol. I, bk. 1, pt. X, p. 194.
16. Ibid.
17. G.R. Elton, "The Triumph of Parliamentary Law", in *The Parliaments of Elizabethan England*, ed. D. Dean, (Oxford 1990), pp. 35–6.
18. P. Zagorin, *A History of Political Thought in the English Revolution*, (London 1954), pp. 189–91.
19. James I, "The True Law of Free Monarchies", in C.H. McIlwain, *The Political Works of James VI and I*, (Harvard 1918), pp. 61–2.
20. John Cowell, *The Interpreter*, (Cambridge 1607), 3R1a.
21. J.P. Somerville, *Politics and Ideology in England 1603–1640*, (London 1986), p. 12.
22. W.K. Jordan, *The Development of Religious Toleration in England*, (London 1936), vol. II, p. 24; W. Notestein, *The House of Commons 1604–1610*, (Yale 1971), p. 133; TNA, SP14/8/70, f. 8–9.
23. W. Notestein, *The House of Commons 1604–1610*, p. 243.
24. *The Parliamentary Diary of Robert Bowyer*, 1606–7, ed. D.H. Willson, (New York 1971), pp. 258–9.
25. W. Notestein, *The House of Commons 1604–1610*, p. 234.
26. John Pym, in *Historical Collections*, J. Rushworth, (London 1659–1701), IV, pp. 201–2.
27. J. Bodin, *The Six Books of the Commonwealth*, ed. M.J. Tooley, (Oxford), bk. I, chap. 8, p. 32.
28. Ibid., bk. I, chap. 10, p. 43.
29. Ibid., bk. I, chap. 8, pp. 31–2.

11 Freedom of Trade as a Developing Principle

The Assertion of Absolute Property Against Prerogative Impositions

One of the defining features of the nation was the development of trade. A growing preoccupation with the conditions of trade was important in establishing a unitary state interest. Parliament was increasingly required to take order for trade by the provision of laws and eventually by the extension of individual rights of consent to the customs. The consolidation of the open, physical land into a political land was associated with a more extended and coordinated structure of commerce, outward and internal. Enclosure and the retreat of the externalized church were not the only changes that undermined the balanced, communal patterns of the medieval world. The scope and character of trade underwent a qualitative transformation from the late fifteenth century.

The new way of business consisted essentially in the development of regional specialization and exchange. This created a dynamic for integration within the kingdom, as well as an enhanced profile for exports overseas. The spread of broad, commercialized marketing came to overlay the more circumscribed context of medieval times, which had been characterized by a multitude of undifferentiated local markets, each providing the same basic range of products for a relatively small area. The new extended market system was also associated with another kind of departure from traditional economic assumptions. With the growing potential of trade, there arose a demand that merchants should be allowed on principle to operate free of impositions and restrictions. This was central to the assertion of a right of "absolute" property, which supplanted the medieval idea of property as "conditional".

The expansion of the scope of trade can be measured on the ground in terms of the number and range of markets. Mark Overton notes the "density" of markets at the beginning of the sixteenth century, when they were still at their greatest in number and most limited in reach. He estimates that most medieval markets served an area of about a ten-mile radius, with little traded outside.[1] W. G. Hoskins was even less generous, suggesting that "in general people found all their earthly needs met within an area of three or four miles at most, within sight of their own church spires".[2] Though perhaps over-restrictive, this captures the essential character of the medieval economy.

Alan Everitt produced a structured analysis of the change that took place during the sixteenth and early seventeenth centuries. He established that "economic concentration is one of the salient themes of inland trade during this period". Everywhere, agricultural traffic was being drawn away from the smaller markets and focused on larger provincial centres like Maidstone, Canterbury, Reading, Newcastle and King's Lynn. These expanded markets became the points of distribution for regional specializations. At the beginning of the sixteenth century, "most towns still served a purely local area, and specialised in no particular commodity". But during the course of the century, "a broad pattern of specialisation becomes clear". By this time, about half of the eight hundred market towns in England were specializing in the marketing of some particular product. The trend to specialization "was clearly marked by the fourth quarter of the sixteenth century, and some of its principal features were already emerging by the 1530s".[3]

This constituted a seminal change in the dimensions of trade and marketing, as each town and region began to concentrate on the more intensive production of particular favoured products of the area, to serve markets and regions further afield. Certain regional marketing characteristics became apparent. The markets of the eastern counties specialized mainly in corn. In the Midlands, the predominant speciality was livestock. The West Country offered another variation, specializing in wool, yarn and cloth. Northern counties showed the highest proportion of concentrated marketing because two specializations were on offer—livestock and cloth.

Commercially minded farmers were thoroughly conversant with the strengths of each regional market, knowing what high-quality goods it offered and when they were at their best.[4] Buyers and sellers travelled long distances to avail themselves of the benefits of regional exchange. The wool market at Doncaster was one of the largest in the kingdom and attracted buyers from more than half a dozen counties and from towns fifty miles away. By Elizabeth's reign, the markets for textiles and livestock served a radius that was generally over twenty miles and not infrequently more than forty. People travelled as far as seventy miles to the great sheep marts of the Midlands and the North. Rotherham market brought sellers from Ellerburn in the Vale of Pickering, seventy miles away, and the sheep fair at Worcester drew sellers from equally distant Camarthenshire.[5] One particular example illustrates the new scope of marketing and the essential impetus behind it. A Westmoreland packhorse man carried cloths from Kendal to Southampton every year from 1492 to 1546.[6] This coincided with the period of maximum expansion of the cloth trade. The regular trip to market, covering almost the entire length of the kingdom, encapsulates how the land was being turned into a commercial unit.

One of the ways in which regional specialization served to increase commercial activity was in enabling each area to concentrate on the production of those commodities to which it was especially well suited but might not previously have been free to exploit effectively. Mark Overton managed to put a figure on the level of absolute advantage gained when two regions began to focus on maximizing their production of the crops or items for which they naturally enjoyed the highest yields. The result was a further increase in overall output of more than half across the two regions.[7] This was roughly the same, we can note, as contemporary estimates of the benefit achieved by enclosing your lands.

An earlier chapter described the diversification and extended productivity generated by the leading example of regional specialization—the growth of sheep farming and cloth manufacture. Since the making of cloth was carried out largely as a cottage industry in the countryside, it could tap underused resources of labour and environment and bring valuable by-employment to areas that were not best suited to productive farming. This was the basis on which the people of the West Riding of Yorkshire were able to alleviate the disadvantages of poor soils and an over-wet climate and to develop a strong economic identity by the rapid expansion of cloth production. In the area around Halifax, which was "planted in the great wastes and moors where the fertility of the ground is not apt to bring forth any corn or good grass", it was said that the people "altogether do live by cloth making".[8] A further consequence was that the manufacturers began to source their wool not as traditionally from the most local market, but from more distant, specialist suppliers. In fact, the people of the West Riding flourished sufficiently by the making of cloth to enable them to import all their other basic needs from the appropriate specializing regions elsewhere: their corn from Lincolnshire, their cheese from Cheshire and Warwickshire, and their black cattle from Lancashire.[9]

The mining industry was another obvious candidate for specialized and intensified production in areas less well suited to arable farming. The extended market in coal became an important element in the development of interregional trade. The southwest corner of Nottinghamshire provides a good example of how this kind of activity could convert a local economy from subsistence to export mode. The area was not blessed with rich agricultural land, but it did possess a resource of convenient coal reserves, and during the sixteenth century it began to exploit them on a substantial basis. By the beginning of the seventeenth century, coal production in the hinterland of Nottingham had reached a level at which the local authorities were happy that they could rely on coal exports to solve any shortage of basic foodstuffs, by means of an annual exchange with Lincolnshire for the grain in which that county was able to specialize.[10]

J.E. Nef estimated that coal output increased fourteen-fold between 1550 and 1680, which was the most significant factor in the "industrial revolution" of the time. The most important centre of coal production, the northeast, generated a flourishing export trade through Newcastle. Much of this was shipped down the coast to the ports of East Anglia, which sent in return their own specialized commodity of grain. The ever bourgeoning market of London was another major customer. It is also notable that by the second half of the sixteenth century, there was a substantial export trade in coal to France, usually in return for salt, and coal from Newcastle was even finding its way to the more distant reaches of the Mediterranean. By 1640, on the basis of this multidimensional trade, England was producing three times as much coal as the rest of Europe put together.[11]

So the development of regional specialization and exchange had a crucial international context. In medieval times, long-distance trade had been mainly in luxuries and very one-dimensional, as described by R. H. Tawney: "With the Mediterranean as its immemorial pivot, expansion had hardly begun. Tapping the wealth of the East by way of the Levant, it resembled, in the rigidity of the limits placed on its commercial strategy, a giant fed through the chink of a wall".[12] In the sixteenth century, these limits were transcended by a flexible, expanding European-wide trading network. Jan De Vries concluded that "medieval commerce did little to alter local self-sufficiency because it remained dominated by long-distance luxury trade". The critical change occurred with the "specialisation pattern", through which the Dutch and the English in particular began to trade basic commodities between regional markets.[13] The cloth trade was England's particular national speciality and by 1513 was being hailed as "one of the most important foundations of trade in the world". Fernand Braudel saw the development of the textile trade as the clearest indicator of the emergence of a capitalist marketing context, displaying "division of labour, rationalisation and specialisation . . . [T]here grew up a textile industry on capitalist lines and connected to distant markets, which was of decisive importance".[14] In the sixteenth and seventeenth centuries, England was in the process of becoming a trading nation in a much more definitive sense than had previously been the case.

The development of internal and interregional exchange in England was encouraged by the fact that inland trade proceeded in a context of exceptional freedom, when compared to other European kingdoms. Unusually, England boasted no large urban centres except London, the uniqueness of which as a coordinating market tended to encourage rather than restrict economic integration. Nor were there any semi-independent provinces. As a result, in contrast to France and Germany, there were few internal tolls or customs barriers between regions in England. "Internal trade went largely untaxed in the period 1500–1700 . . . [O]ne of the principal strengths of the network of internal trading . . . was the relative lightness of passage tolls, pontages and other duties on domestic trade which created in much of Western Europe a fairly rigid series of customs unions".[15] Even the tolls that had once existed in English town markets had largely fallen into disuse by the sixteenth century. An increasing amount of business was in any case being done outside the confines of the towns, often at inns. These trends towards private marketing "seem to have developed very substantially in this period, as functions of the spread of the major interregional trades in agricultural commodities".[16]

At the same time, it was becoming more difficult to apply the traditional moral restraints by which the authorities had sought to ensure that supplies and prices were within the reach of every pocket. The rules that had once been normative were now enforced only at times of severe dearth, and "regulation gradually broke down" under the pressure of the sharp increase in market activity. "The Elizabethan government came to accept that the grain market was driven by the desire for private gain".[17] The contemporary William Harrison regretted the passing of the moral economy. He looked askance on the fact that sellers travelled great distances to particular markets, merely "to seek the highest prices". He was critical of the general failure to apply the assize of bread, which was meant to set the quality and price of that basic commodity. And he was dismayed that in most markets the standard of goods was not "any whit looked into, but each one suffered to sell or set up whatever he himself listeth".[18] The assumed freedom to do "whatever he himself listeth" was indeed the force that was undermining the principles of moral economy and conditional property.

Freedom of Trade as a Developing Principle

The effect of the relatively unrestricted trading conditions in England was not just the physical encouragement of commercial activity. It also contributed significantly to the changing economic psychology that was supplanting the old moral codes. The very existence of the context of freedom and the benefits that it seemed to produce engendered the thought that it should be extended as a matter of principle. So the idea of freedom of trade, together with the enhanced status of the concept of consent in England, led to the assertion that English people had an actual right to trade at liberty without being subject to the arbitrary exactions and monopoly grants in which the authorities were all too likely to indulge.

The term "freedom of trade" carries the impeccable authority of Sir Edward Coke, who employed it on occasion and became a powerful proponent of the principle. A philosophy of freedom of trade can first be seen emerging in the mid-sixteenth century. It appeared in the most incisive of Tudor economic analyses, *A Discourse of the Commonweal* of 1549. One of the author's principal concerns was the arbitrary enclosure of the commons and the loss of arable land, which was regarded as having taken place at an unacceptable level in recent decades. Punitive legislation had regularly been attempted "yet small remedy found that took effect". The authoritative voice of the Doctor proposed that a more promising way forward was to accept the irresistible power of the profit motive and to "make the profit of the plough to be as good rate for rate as the profit of the grazier". The way to do this, he said, was to give the products of arable husbandry the same level of freedom as heretofore enjoyed by the graziers. "That the husbandman might have as much liberty at all times to sell his corn either within or without the realm as the grazier hath to sell his". The rules that restricted the export of corn when the price was high should be relaxed. The *Discourse* reflected a belief that the best way to "cherish" the production of corn was now by "allurements and rewards". "Take these rewards from them . . . what men will plough or dig the ground, and who will venture overseas for any merchandise?" For the first time, freedom of trade was being put forward as a general principle. The *Discourse* suggested that punitive impositions always did harm and that the cause of the downturn in trade had been the government's raising the custom on cloth and placing a surcharge on it in the last subsidy, "which was the very highway to make the clothiers give up their occupation".[19]

This was the basic rationale of freedom of trade, asserting that the best way to increase trade and production was by allowing maximum profits and that the surest way to undermine it was by arbitrary impositions. Another tract published at the same time as the *Discourse* repeated the prescription—to raise the level of incentive for arable farming by setting the corn trade free. "To suffer all manner of corn to be freely transported over the seas as well as any other kind of merchandise".[20] The proposition immediately found echo in some letters sent to William Cecil in the Privy Council. In December 1550, Sir John Mason wrote protesting against the restrictions being placed on the export of cheese and butter. He pointed out that such restraints did not make things cheaper—they simply depressed production. "For who will keep a cow that cannot sell the milk for as much as the merchant and he will agree upon?"[21] In the following year, Cecil received a letter from William Lane, extending exactly the same advice in respect of the rather different but important products of the mining industry.[22]

In March 1559, the experienced voice of John Hales joined the chorus, with a plea on behalf of the most important of English exports—woollen cloth. He reiterated the freedom of trade rationale. Wealth production was ensured by the liberty to profit. Restrictions and arbitrary exactions were counterproductive. The proposed imposition on cloth was not the way to increase royal income "but rather a means to make no English merchants". It would undermine the kingdom's most essential trade and manufacture. A more efficient balance in economic affairs would be achieved by allowing the maximum freedom to profit in the market. "For merchants trade for gain, and when it ceases, they trade no more".[23]

Most of these pleas recognized that their advice ran counter to the official view and was unlikely to be heeded. It did indeed long remain the government's standard procedure to regulate trade through restrictions and impositions, as well as grants of privilege. Eventually, the developing assumptions of freedom of trade came into direct confrontation with the policies of the crown and led to the first formal challenges to aspects of the royal prerogative. Some of the earliest assertions of freedom of trade as an inherited right came as a reaction to the government's increasing tendency to issue monopolies. Thus in 1586 the Boston salt traders protested against the salt monopoly. They found it unacceptable that outsiders should be given privileged control of the trade and that they themselves could "not convert their salt pits and salt coteries being their inheritance to the best use". They suggested that it was "contrary to the laws of this realm".[24] This was not really so, but it was becoming the settled view of the nation's representatives that it ought to be so. The cause of the Boston traders was duly taken up in the Commons. The salt monopoly would "walk in the forefront of abuses" when parliament came to address the grievance of monopolies in 1597 and 1601. It was said to be costing the ports of Boston, Lynn and Hull £3,000 a year.[25]

The radical force of the emerging principle was strikingly illustrated in an anonymous tract of the late 1580s, asserting the right of freedom of trade at the most general level. The writer condemned all monopolies on the basis that they were the engrossing of trade into the hands of one or a few, "which ought to be free and common to all the citizens of the said commonwealth". This was the cutting edge of freedom of trade, allowing no privileges and no exceptions. No corporation should be allowed to exclude everyone else from a trade. It was "against common right and the law, under which we are born, and is our inheritance".[26] The right of freedom of trade was being advanced as a generalized belief.

The broad sentiment behind this conviction became clear in 1604 when the Commons introduced a free trade bill, which would have given legislative force to the same uncompromising stance. It proposed "free liberty to trade into all countries". The intention was to break the restrictive monopoly powers of the Merchant Adventurers and, for good measure, the powers exercised by the Russia, Eastland, East India and Turkey Companies in their respective trading spheres. The bill sought to make it lawful for any of the king's subjects to have free passage and trade into any country, and declared that any grant of a privilege that tended to create a monopoly should be void. It is something of an understatement to say that this was "to some degree an assault on the government, which had favoured the great companies".[27] In fact, it was the assertion of a general liberty from imposed restrictions, and this was the platform from which a series of challenges to the prerogatives of the crown would be launched. It was notable that the initiative came from the general body of merchants themselves. The Venetian ambassador recorded that "the Lower House, in a petition complaining of monopolies signed by many merchants, has dissolved all companies".[28] That might indeed have been the effect if the bill had taken full force. But although it passed the Commons "by a sweeping vote", it was obstructed by the government.[29] It retains great significance as the first indication that the merchants and their representatives in the House of Commons were prepared to challenge the position of the crown in the name of freedom of trade. It underlines the fact that the subjects' claim to an "absolute property" in their possessions was advanced in the context of trading rights. The expressed justification for the bill was the first public statement of that principle. "All subjects are born inheritable, as to their lands, so also to the exercise of their industry in those trades . . . [I]t is against the natural right and liberty of the subjects of England to restrict it to some few".[30]

The radical force of this proposition was underlined a decade later in John Kayll's book *The Trades Increase*. Kayll would also have "dissolved all companies" on the principle of an inherited right of freedom of trade for all. Much of the tract is devoted to a somewhat rambling complaint about the perceived reduction in the nation's shipping, bemoaning the restrictions and discouragements placed on it. Although this sounded vague and impressionistic, it was quite prophetic, for the "decay of trade" was imminent. Kayll's proposed solution was forceful and coherent. "Let me, without offence, propose the consideration of a remedy thereto, even by a freedom of traffic for all his majesties subjects to all places".[31] Kayll was aware that this was a contradiction of the official stance. He was advancing an uncompromising vision, but it was in a direct line of descent from the radical pamphlet of the late 1580s and the Commons' very assertive free trade bill of 1604.

Kayll proposed an end to all privileged companies, to let "the whole corporation of merchants reap comfort, in that they may communicate with all adventures, and the universal body of the subjects of the land content in that they may become merchants". He then offered a powerful conception of this freedom as a general right and liberty of the kingdom. Company restrictions were unacceptable "to fellow subjects and equal citizens in this great monarchy, to be so serviceably tied and subject one to the other, and the rather for that those privileges, by the indulgence of the Prince being granted as a reward to some . . . are strictly used to the eternal benefit of a few and the wrong of all the residue".[32] This illustrates how closely the demand for freedom of trade was bound up with the developing assertion of an "absolute" right of the individual not to be subject to arbitrary impositions and restraints from any other quarter. The tract contradicted the position of the government at every level, and it was not surprising that Kayll found himself cooling his heels in prison. He had not merely condemned the system of granting company privileges as a policy—he had declared it to be invalid, as a transgression of the independent rights of the subject. This was the assumption that King James always found most threatening and was regularly obliged to suppress. It was particularly challenging when the radical claim of inherited rights was applied to areas as actively contentious as freedom of trade.

This fundamental divergence of approach became set. The early Stuart government never freed itself of dependence on the granting of patents of monopoly as a means of regulation and patronage. Nor did it manage to dissuade the Commons from persistently asserting a liberty from arbitrary restraints and exactions. The opposition to monopolies and the rejection of imposed customs dues were intimately connected. They reflected the same economic and constitutional energy as two aspects of the drive for freedom of trade, and they had a common enemy in the various forms of government and administration by patent. In consequence, the freedom of trade perspective was asserted broadly and continuously and almost always in contradiction of government policy. I have described this elsewhere as an increasingly generalized polarity between "parliament" and "patent". There were arising two sharply differing perspectives on preferred methods of public finance and administration, and they were constantly in play.

Between 1614 and 1620, with parliament in abeyance, the king and court had indulged in what can be described as an unrestricted binge of patents of monopoly, the incidence of which more than quadrupled. "The world doth even groan under the burden of these perpetual patents . . . now said to have multiplied by many scores".[33] This became a problem when James was obliged to call parliament to help address the foreign policy crisis in 1620. To encourage a mood of cooperation, the king found it expedient to promise that he would listen to specific grievances against particular monopolies. Many modern historians, in their determination to minimize the assertiveness of parliament, have allowed the impression to grow that the government was somehow taking the initiative in the attack on monopolies and had suddenly swung into opposition against its own standard policy. This is one of the most damaging distortions contrived by revisionist history. If some figures in the Privy Council chose at certain stages to join the popular agitation, it was usually to divert blame from themselves or undermine their governmental rivals. It did not alter the fact that the strong inclination of early Stuart regimes was to favour and depend upon monopolies as an instrument of regulation and patronage.

The Commons, on the other hand, sought to bring the practice to an end in its present form as an unacceptable breach of the subjects' assumed right to freedom of trade. The general bill to outlaw monopolies was seen as the most widely desired of the many bills that were brought forward by the House in 1621. It was, said Sir Nathaniel Rich, the measure on which the heart of the country was set.[34] The aim was to wrest control of monopoly patents from the sphere of the prerogative, reflecting the intent of the Commons' original legislative proposal on the matter in 1601. It was introduced on 5 March 1621 by Sir Dudley Digges and Sir John Walter. The latter summed up the purpose: "Let all patents be called in and suppressed, and have none granted hereafter. Let all projectors henceforth incur praemunire for preferring suits".[35] Walter suggested that if they could get this measure through, it would solve the problem in perpetuity, and they could avoid provoking the king by the further pursuit of ministerial referees. S.R. Gardiner agreed: "It was of greater importance to define the law . . . [I]f they could convert into law the bill before them, it would never again be in the power of any minister, however high in favour, to divert disputes relating to the commercial privileges from the ordinary courts".[36] The bill fell victim to its own radical force. In undermining the king's power to sustain the system of monopoly patents, it would be, as McIlwain said, "the first statutory invasion of the royal prerogative".[37] It failed to clear the House of Lords.

The divergence over the monopolies bill was one manifestation of a broad, underlying dichotomy of approach to economic regulation. This was illustrated by the fact that although the government took a persistently negative view of free trade legislation, the Commons became ever more committed to the concept. They regarded it as a public priority and felt some resentment that they were left to take order for it. As they contemplated a programme of free trade bills in 1621, William Nyell noted that the only freedom they currently enjoyed was of trade to France and Spain, under the act of 1606, "and that by care of this House . . . For they were both under patents".[38]

In 1621, free trade legislation became in effect the economic policy of the Commons, as their principal response to the critical problem of the "decay of trade". As John Kayll had foreseen in 1615, a chronic trading depression had set in. One cause was the ill-fated Cockayne project, which had succeeded only in provoking Dutch reprisals. There was also the disruption of the German and Baltic markets, which had been important for English cloth exports. The Commons' solution echoed what Kayll himself had recommended—to encourage free trade wherever possible. Sir Edwin Sandys suggested that the best cure would be to set up seven committees to challenge the seven great monopoly companies of London. This neatly symmetrical proposal illustrated the breadth of the Commons' intent. They brought forward a raft of free trade measures in an attempt to address the depression. As so often, it was Sir Edward Coke who summed up the rationale. "Freedom of trade is the cause that the Low Countries so prosper. They are not burdened with impositions to burden trade, nor monopolies to restrain it".[39]

So a series of bills were proposed with the aim of establishing this freedom in every aspect of the vital cloth trade.[40] There was an early move to address the downturn in the trade by a bill for "free liberty of the buying and selling of wools". It was noted that less than half the amount of cloth was being produced as formerly, and many clothiers and cloth makers were out of work. It was thought that the monopoly of the Merchant Adventurers was restricting the market and that they should be deprived of the power to decide which cloth should be sold and which not. The solution was a bill for free trade.[41] This was later reinforced by a bill for "restoring the free trade of the merchants of the staple for the export of cloth into parts beyond the seas".[42] This too was aimed at undermining the monopoly of the Merchant Adventurers and empowering other merchants to export cloth independently from the outports. MPs certainly believed that a major cause of the depression in the nation's crucial cloth-producing enterprise was the artificial restraints placed upon it, and that the answer was to create conditions of free trade.

There was an attempt to deal with a specific problem hampering the Welsh cloth trade, which was believed to have flourished in conditions of freedom but was now suffering from restraints imposed in some English markets. A bill was therefore proposed for "the free trade and traffic of Welsh cloths, cottons, linens, etc., through the kingdom of England". Other measures were also aimed at providing free trading rights for the principality. A week later came a bill for "free trade in Welsh butter".[43]

The programme had a national reach, and the northeast was not neglected. Another bill brought forward early in the session was "for the free traffic of sea-coals in and out of the counties of Durham and Northumberland".[44] For good measure, there were bills to confirm free trade into specific overseas markets, and a general bill "for liberty of trade into all countries".[45]

To complete its all-embracing scope, the freedom of trade perspective also covered the agrarian sphere. It was not just that many of the free trade bills concerned agricultural products or items manufactured from them but, more specifically, that this parliament took an important step in changing the relationship between legislation and the process of enclosure. A previous chapter described how the dearth of 1597 had induced the last acts seeking to restrain enclosure as a cause of depopulation. Thereafter, the flow of legislative initiative gradually switched direction towards a positive view of the process. The year 1608 had seen the first limited pro-enclosure measure. In 1621 came the first general bill with the aim of advancing the practice. In 1624, the existing statutes against enclosure would be repealed.[46] The turning of the legislative tide in favour of enclosure was an integral part of the growing assertion of freedom of trade. It was another aspect of the core demand that each individual should be free to use his property "as he himself listeth".

Historians have sometimes sought to diminish the significance of the movement for free trade in the early seventeenth century on the basis that it was a sectional protest of some of the merchants of the outports against the London monopolies, rather than the fully fledged, generalized concept that emerged in the later eighteenth century. This is to conceal the genuinely radical and national force of the earlier manifestation of freedom of trade. The combined animosity of the outports against the privileged elites of London did in any case constitute a national perspective. And in fact the wider body of merchants in the capital itself were also committed to the furtherance of freedom of trade against arbitrary restrictions and impositions. Furthermore, this was a principled stance shared by merchants, gentlemen landowners and commercializing farmers alike. These close but broad connections are demonstrated by the ascendancy that the movement held in the economic policy of the House of Commons from 1600 to 1640.

In truth, the freedom of trade concept could scarcely have been more national in scope. This is illustrated by exchanges in the House in 1621 over yet another free trade bill—for free fishing off the coasts of Newfoundland. It was to counter a patent granted to Sir Ferdinando Gorges in 1620 in order to control the activities of the West Country fishing fleets, ostensibly in the interests of the New England colony. The West Country merchants and fishermen saw this as just a form of tax on their very lucrative trade. Consequently, the free fishing bill was proposed in every parliament of the 1620s by either Dartmouth or Plymouth. William Nyell, from Dartmouth, declared that the colonists could "take the first place now, if they can get it".[47] This was the voice of a genuine, self-confident free trader and economic individualist, and it commanded broad support in the Commons.

This included the powerful figure of Sir Edwin Sandys, who represented the Cinque Ports, with no direct interest in the cod fishing but with a firm commitment to the broad struggle for freedom of trade. Drawing on his own knowledge of the colonial sphere, Sandys assured the House that the colonists had no need for this kind of privileged protection—the fishing should be free for all.[48] The West Country MPs later repaid Sandys for his support by giving their vocal backing to the protest of the Cinque Ports against the most generally disliked of all monopolies—that of the Merchant Adventurers in the cloth trade. The national perspective was underlined by the contribution of another prominent free trader, Sir Edward Coke, who stated that there was "no greater cause in all this parliament" than the Cinque Ports petition against the Merchant Adventurers.[49]

To the royal government, the movement for freedom of trade merely produced unwanted initiatives that threatened to undermine the crown's ability to regulate and tax the mercantile sphere. The Commons' bills often seemed to take this form, and this was one reason why James I ceased to take a sympathetic interest in the legislative process. We will never know what effect the programme of free trade measures of 1621 might have had in alleviating the depression because the king chose to dismiss the assembly before any of them could reach the statute book. Sir Nathaniel Rich, a leading proponent of the bill against monopolies, was distraught at the loss. "Our estate is a decaying estate. We were preparing remedies. We are taken off and interrupted".[50] This angry reaction, in effect questioning the king's decision, reflected the strength of the Commons' belief in the necessity of their policy, both in practice and in principle. They had been attempting to assume responsibility for the national economy and solve its current problems. They had sought to address the depression by creating a general context of free trade, not merely because they regarded this as the appropriate practical solution but also because they now saw freedom of trade as an essential right.

An interesting echo of the breadth of the free trade perspective came from a merchant whose official brief was to advocate restraint and to resist the more uncompromising elements in the movement. Gerard de Malynes was an authoritative voice and sometime adviser to the government. He is normally regarded as attributing the decay of trade to the vagaries of the exchange rate and asserting the need for an orderly trading context, as against complete freedom. Yet the most striking passage in his book is a condemnation of the monopoly of the Merchant Adventurers. One of the main causes of the depression, he said, was the restrictive practice of the great privileged companies. He pointed the finger at the Eastland Company, the Russia Company and especially the Merchant Adventurers. "Having engrossed into their hands . . . the sole power of exporting all white cloths, all coloured cloths, kersies, bays, says, serges . . . hath abated the trade".[51] This undermined other merchants "from all the staple ports at London, Westminster, Bristol, Southampton, Royston and Newcastle that heretofore exported either cloth or wool or both, which now they may not".[52] This monopoly was "a general prejudice to the whole kingdom . . . has caused our home trades to decay, our manufactures to decrease, and our commodities to lie upon our hands unsold". "Shall this be proclaimed a free trade, when we ourselves are in bondage?"[53] De Malynes was clearly prepared to go a long way to accept that freedom of trade was a right to be pursued and to echo the thoughts of the Commons that the absence of it was a major cause of the depression.

The parliament of 1624 did get an opportunity to pass some legislation. This was an indirect result of the attempt by Prince Charles and the Duke of Buckingham to bring together the Commons and some usually fractious and mutually hostile peers in a grand alliance to try to persuade the king to reverse his pro-Spanish policy. In the event, James managed to resist the pressure. But parliament was at least allowed, for the only occasion between 1610 and 1640, to get a significant programme of bills onto the statute book. In the relatively favourable climate, the Commons made some headway in changing the balance of public provision in favour of the freedoms that they desired. The opportunity was taken to phase out some of the most ancient traditional restraints on the flow of trade. The statute of 1361 forbidding the export of corn until a license had been obtained from the crown was repealed in 1624.[54] Other long-standing measures for protecting the supply of corn and preserving the practice of tillage and husbandry against enclosure were also repealed.[55]

The most prominent new provision in the direction of free trade was the monopolies bill. In 1621, it had come to grief in the House of Lords, where it was seen as an infringement of the royal prerogative. In 1624, the agreement of the Lords ought to have been easier to achieve, but even now they made continual objections aimed at protecting either their own particular privileges or the rights of the crown. In the end, the Lords and the king gave way under pressure from Charles and Buckingham, who understood that the passage of the bill was a condition of the Commons' cooperation in other fields. So the monopolies bill passed into law, "in its key passages almost word for word the text of the one rejected by the Lords in 1621".[56] It foreclosed the exercise of the prerogative in this area and replaced it with the supervisory control of parliament and the Common Law. It was "the first statutory invasion of the royal prerogative".[57]

The constitutional significance of the act has sometimes been minimized on the grounds that it provided for some exemptions. It contained no bar to granting privileges to companies and corporations. But the Commons were not thereby compromising on the core principle of opposition to monopolies. At this time, there was a pragmatic need for companies to be given protective privileges to enable them to establish a base for foreign trade. In the 1650s, the state would begin to assume direct responsibility in this field, but until that time it was difficult to avoid delegating executive powers to the companies. In the 1630s, the Caroline administration was able to exploit the exemptions and grant monopoly rights under the guise of corporations, in return for annual sums paid to the crown. But in a sense this simply underlined the dichotomy. The early Stuart kings never managed to free themselves of dependence on regulation by patent and the grant of monopoly privileges. Parliament on the other hand, remained determined to classify these practices as essentially illegitimate. As Coke made clear, the crucial provision in the act was that there should be no monopoly that created "any liberty of sole buying and selling".[58]

Other particular initiatives for freedom of trade were also revived in 1624, though with less success. The free fishing bill was brought forward again, as always. But even in this parliament, where there was a presumption of cooperation, the royal government was not prepared to approve it. There were prolonged discussions, which ensured that the bill was sent up to the Upper House "late enough for the Lords to claim the excuse of lack of time to pass it".[59] Most interestingly, the continual reappearance of the free fishing bill shows the determination of these boroughs to keep pressing for the conditions of freedom of trade that they had come to regard as vital to their livelihood and their prosperity. It also reflects their confidence that the Commons at least could be relied upon to support them. And in fact, in the relatively favourable climate of 1624, when the Commons had the wind behind them, they did find other ways of addressing the problem, with some effect. The powerful committee of grievances, chaired by Sir Edward Coke, with a general license to control the activities of monopolists, pressed the principal patentee, Sir Ferdinando Gorges, to accept that notwithstanding the powers of restraint in his grant, he would hereafter allow the fishermen to "freely visit the coasts of Newfoundland and fish there without any interruptions". Having forced this concession, the Commons seized the moment and declared free trade in the area. The merchants of Plymouth gratefully preserved a copy of the resolution:

> Thereupon it is resolved by the same House of Commons that the fishermen of England shall have free liberty of fishing on the said coasts of New England, with all the incidents necessary of drying their nets and salting and packing their fish. And it is secondly resolved that in the opinion of this House the clause in the said patent of confiscation of the ships and goods is void and against law.

The Plymouth representative added the note, "which said order is remaining in the town chest".[60] They were keeping the resolution as a statement of the legitimacy of the conditions of freedom that they had come to see as crucially important to their trade. In effect, they were happy to regard parliament as the arbiter of the law in this field and to accept it as an authoritative, independent voice, even though it might be contradicting that of the crown. This revealed a basis on which the merchants might be prepared to envisage an enhanced constitutional status for the representative assembly. As we shall see ahead, the merchants of Dartmouth embraced exactly the same kind of initiative in pursuance of the other main provision of the freedom of trade agenda—the demand for the surrender of prerogative impositions.

The Assertion of a Right of Absolute Property against Impositions

The dispute between prerogative and representative finance turned most of all on the question of impositions, which was indeed the subject of the first full-blown constitutional confrontation between crown and Commons when, in 1610, MPs launched an open assault on the monarch's power of imposing customs dues. In essence, they were asserting freedom of trade against arbitrary exactions.

In 1606, the London merchant John Bate had refused to pay the imposition on imported currants. He was not a lone voice. The initiative was widely supported by merchants in the capital.[61] Parliament too was already concerned with the issue. Impositions on currants and other commodities, together with complaints against the raising of customs generally and objections to a wide range of monopolies, were the main items on the Commons' list of grievances, when the case of John Bate was added to them. Despite official obstruction, the Commons determined that the merchants should be heard, and they stated their case vociferously. One threatened indeed "to leave all rather than this shall stand—go beyond seas".[62] This was the core freedom of trade argument, to which John Hales had given early expression in 1559. Impositions were, "a mean to make no English merchants".[63]

The king referred the matter to the Court of Exchequer, which unanimously concluded that he had the right to levy customs by prerogative. Baron Fleming added the fateful gloss that the monarch did this by virtue of the discretionary authority that he was free to exercise as guardian of the general public good, and the regulation of trade came under that heading. Robert Cecil, who had taken on the unenviable task of managing the treasury, felt it safe to exploit the Exchequer decision and imposed duties on a wide range of commodities. The principle of freedom of trade, far from being advanced as the merchants intended, had suffered a substantial setback. In fact, Fleming's statement threatened to place the customs dues in a realm of "absolute" power, outside the representative sphere. This was the opposite of parliamentary projections and would have to be rebutted. So when the House reassembled in 1610, there ensued the first outright constitutional clash between king and Commons—the impositions dispute.

An indication of the real force of the Commons' ambition came in Lord Treasurer Salisbury's opening contributions to the session. He clearly had some apprehension of a developing challenge to the royal power of impositions, for he found it necessary to state that there was no room for movement in that respect, since "in some things the prerogative was inherent and inseparable, as in impositions".[64] On the other hand, although the rights of wardship were also part of the prerogative, they were not, apparently, an inseparable part. The provision of wardship might be made less onerous or even abolished altogether if the Commons were to offer the crown a permanent income instead.[65] In other words, wardship might be up for negotiation, but the king's power of impositions was not. The Commons' response confirmed where their priority lay. An end to wardship was certainly sought by many landed gentlemen, but the deal was never done. The Commons' main ambition was to extend the range of rights of consent, in this instance by launching a full-scale attack on the king's prerogative of impositions, though they had been told, quite specifically, that this was out of bounds. It is clear that the attempt to undermine the power of impositions was an imperative for them. In common with the broad classes of merchants and yeoman freeholders whose interests they represented, MPs were primarily concerned to establish the principle of freedom of trade by denying the right of the king to set customs dues by discretion. So as already suggested, the development of an "absolute" property right was less about changes in tenure than about the assertion of a freedom to trade and use property without arbitrary restrictions and exactions. It was in this context that the claim of absolute property was now explicitly advanced.

When parliament reassembled in 1610, the Commons ordered a search of the records to assess whether it might be possible for them to challenge the decision in Bate's case. The king attempted to nip this in the bud with an instruction that they should not presume to query his prerogative or propose laws against it. It was not their place, he said, "to question what a king may do".[66] But by trying to pre-empt the attack, James had escalated it, for the Commons could now assert that their liberty of free speech was being denied. In a sense, the king had strengthened their hand, for although they had no right to attack the prerogative of impositions, they did in certain respects have a right of free speech, which they now asserted.

There ensued a dispute about what exactly the Commons had the right to discuss. It sounded like an exercise in hair-splitting, but it reflected an important development. Solicitor General Sir Francis Bacon said that they should restrict themselves to practical and individual grievances, for they were entitled to discuss "the right or interest of any subject or the commonwealth". He was using the term "commonwealth" in its older sense of the welfare of the populace and trying to limit the Commons to their traditional sphere of the pragmatic and particular, unconnected with general issues of authority or political principle. MPs pointed out, however, that the power of impositions was now seen as affecting everyone's rights. "If it be true as Mr Solicitor confessed that the parliament may not be inhibited to debate of anything which concerns the rights of particular subjects, much less may they be inhibited in this matter of impositions, which concerns the rights and interests of all the subjects in general in those things which they enjoy".[67] This revealed a crucial transition. The issues that the Commons addressed were ceasing to be just a multitude of local interests and were coming to be perceived as overall public imperatives.

In the interests of his prerogative, James would have been well advised to stand by his original intention not to allow the Commons to pursue the matter. But in the statement asserting their right of free speech, they had taken care to hint that if they were given permission to discuss it, this would encourage them to loosen the parliamentary purse strings. So the king relented and allowed a debate, while underlining the short-sightedness of this decision by reiterating that they did not have the right to question the power of impositions as such but only its particular practical inconveniences.

For the Commons, the whole purpose was to question the king's discretionary powers in this area, and when the great impositions debate began at the end of May 1610, a series of parliamentary speakers stepped forward to declare that it was unlawful for the king to raise customs dues by prerogative and that it ought always to be done by consent in parliament. In terms of the content of the law, their case was not strong. They cited medieval precedents that for certain times and purposes, kings had agreed that customs should be raised by consent. But this had never become a general principle, and it did not alter the orthodox constitutional position that the king could raise taxes by prerogative when he judged it necessary for the public safety, as recently affirmed by the Barons of the Exchequer. The real strength of the Commons' case was political. It lay partly in the force of their determination, as representatives of the kingdom, to establish a comprehensive right of consent in the raising of the customs. It also lay in the fact that there now existed a concept of sovereign lawmaking authority that gave them a rationale for displacing the prerogative.

This was set out in the opening of the first parliamentary speech, by one of the most vociferous of free traders, Nicholas Fuller. He offered a logical syllogism: In England, the laws could not be made or altered without consent; freedom from impositions had once been a law; and therefore this freedom could not be taken away without consent.[68] Royal supporters attacked the premise of the syllogism because they feared being wrong-footed by Fuller's attempt to direct the issue away from the content of the law and towards the form in which it was made. But another major parliamentarian voice, James Whitelocke, came to Fuller's support.

> It hath been alleged that those which in this case have enforced their reasons from this maxim of ours 'that the king cannot alter the law' have diverted the question. I say under favour they have not; for that in effect is the very question in hand; for if he alone out of parliament may impose, he altereth the law of England. So in this point the question is, whether the King's patent hath the power of the law, or not, for if it be not maintained that he hath, it can never be concluded that he can transfer the property of his subjects' goods to himself, without the assent of them.[69]

The intent was clear. Like Fuller, Whitelocke was seeking to bring the raising of the customs under the same rights of representative consent that applied to the making of sovereign law. In effect, they were saying that if the king raised customs by prerogative, it was *as if* he were making law by patent, and since in England the king's patent no longer carried the force of law, this could not be valid. The same argument underlay Whitelocke's more famous contribution to the debate, in which he outlined the idea that the king had two levels of power—one that he exercised by his own authority and another that he could only exercise in parliament. The killer punch was that since it was only in parliament that the king could perform the sovereign functions of making law and raising taxes, therefore it was the second power that was "the greater of the two and doth rule and control the other".[70] Here was the explicit invocation of the radical potential entailed in the theory of sovereign representative law—the notion that the "absolute" power of lawmaking could only be exercised by parliamentary consent and not by the king alone. In 1610, Fuller and Whitelocke were employing this principle to undermine the king's prerogative in another sphere—the raising of the customs dues. All such sovereign functions, they suggested, should now be done by consent in parliament.

So the parliamentary agenda in 1610 was specifically to incorporate the raising of the customs within the sphere of representative consent. Historians have been wrong to suppose that the battle against impositions was undertaken as a surrogate defence of the principle of consent in land taxation.[71] On the contrary, it is clear that the express aim in 1610 was to extend that right to the customs dues. This was stated just as plainly as it had been in respect of the Free Trade bill in 1604. People were "inheritable" in the property of their trades, just as in the property of their land. Nicholas Fuller's summary was unequivocal: "the subjects have such property in their lands and goods as that without their consent the king can take no part thereof from them lawfully by any kind of imposition or other wise without an act of parliament . . . that the laws of England do as tenderly preserve the right and liberty of the subjects in their lawful and free trades, mysteries, and manual occupations as in their lands and goods".[72]

Fuller waxed lyrical about the value of commercial activity. "For no trade in England ought more to be cherished than the trade of merchandise, which doth greatly uphold the wealth, strength and credit of the land, and doth vent whatsoever the fruitful labour of any subject doth yield".[73] This reflected an awareness of the growing scale and substance of the nation's trading activity, which undoubtedly concentrated minds on the importance of controlling the rewards. And so, concluded Fuller, the law should "more tenderly preserve the subject's freedom of his trade (since by trades and occupations commonwealths are upholden) than the inheritance of his lands".[74] A supporting comment of William Hakewill's illustrated how the concept of absolute property was being generalized in the context of trading freedoms and the growing significance of the commercial sphere. Trade now went on in every dimension, and it went on all the time. He pointed out that if subsidies "which came but seldom were brought to certainty, then much more custom might, which hapneth daily".[75] The words of Sir Heneage Finch showed exactly how this translated into a theory of absolute property. "Every man hath a property in his goods against the king as against all other men . . . and any man that would take away any man's goods must show his reason for it".[76] It was the generalization of the principle that goods and property could only be taken with consent.

Thus the claim of absolute property was first explicitly advanced in the context of extending representative rights to the raising of the customs. The right of consent was being stated as a general principle because its growing public force had generated a desire for it to be applied more broadly. The essence of this development was encapsulated by Thomas Hobbes, who perceived the emergence of a new economic ethic, particularly prevalent among the citizens and traders of market towns and entailing the belief that each man was, "so much master of whatsoever he possessed, that it could not be taken from him upon any pretence of common safety without his own consent".[77] This was a precise description of the position adopted by the Commons in 1610. It was now assumed as an inherited right of the subjects and an essential condition in which they wished to work and trade.

The Commons' espousal of the merchants' cause illustrates the supposition of earlier chapters that the emergence of absolute property in England rested on two factors: the developing claim of freedom of trade and the growing status of the concept of consent as embodied in parliament. The close combination of merchants and gentry is reflected in the fact that most of the detail of the speeches just referred to comes from a record of the parliamentary debate preserved by the merchants of Dartmouth. They had developed a constructive relationship with a leading Commons free trader, the gentleman lawyer Nicholas Fuller. They acquired copies of his draft speeches, as well as compiling a rough set of notes of their own of other speeches that challenged the royal prerogative of impositions.[78] This showed that they were now taking a constitutional interest in parliament as the best hope of promoting the principle of freedom of trade, which they had come to regard as the main condition of economic success. It has already been noted that while taking over the bulk of the borough seats in the Commons, the gentry had also taken on a commitment to the interests of the mercantile sphere. This reciprocation appears to have been distinctive to the English context, just like the exceptional force of the concept of consent and the assumption of freedom of trade itself.[79] From these special connections emerged the notion of absolute property.

The capacity for radicalism among the merchants has been greatly underrated by historians of all persuasions. The tendency among recent academics has been to stress the formal aspects of the urban structure that can be interpreted in a conservative light. Thus a study of Bristol concluded that, notwithstanding their worldwide trading interests, the merchant elite was an exclusive group with a conventional and hierarchical concept of authority.[80] Such hypothetical assumptions overlook the very marked oppositional stance that the merchant body adopted in practice, initiating the struggle against prerogative impositions and trading restraints, and providing the most substantial support for parliament in the Civil War. Even some Marxist historians have been reluctant to acknowledge the powerful dimension of liberal freedoms generated by the development of commercialized, interregional marketing. Maurice Dobb regarded the mercantile elites as being in collaboration with the old feudal order.[81] Similarly, Robert Brenner has suggested that in the first four decades of the seventeenth century, the London merchant community was essentially passive in its relationship with royal government.[82] Even for the late 1620s, when the mercantile protests are too loud to ignore, Brenner supposes that the real opposition was coming from a radicalized House of Commons, turning "the merchants' simple economic grievances into a constitutional issue".[83] In truth, as we have seen, the evidence of the direct parliamentary involvement of the Dartmouth merchants shows that, already by 1610, they were placing their economic interests in a strongly constitutional light. And in fact it is quite clear that at crucial moments in 1604 and 1606, the merchants generally were shaping the attack on monopolies and impositions. Again, as regards the mid-1620s, Brenner seems to be contradicted by his own evidence, which indicates that it was the merchants who were taking the lead in reviving the challenge to prerogative impositions.[84]

The prominent radicalism of the mercantile class from 1604 onward must be clearly recognized if we are to understand the force of the freedom of trade agenda that was at the heart of the opposition to the crown from the turn of the century through to the revolutionary measures of early 1641. The foregoing pages have illustrated the basis and substance of merchant radicalism in the assertion of a right of freedom of trade. Christopher Hill produced a persuasive summary of how this worked in context. He noted that writers like Raleigh and Bacon "summed up the ways in which traditional authority was being undermined . . . caught the optimism of the merchants and artisans, confident in their new found ability to control their environment, including the social and political environment, and their contempt for the old scholasticism".[85] This captures the self-confidence and assertiveness of the merchants and establishes their relationship to other crucial developments of the time, as in conjunction with the gentry and the yeoman farmers, they broke away from the restraints of the medieval past in the pursuit of liberal freedoms.

Following the constitutional confrontation of 1610, the Commons' desire to overturn the right of prerogative customs dues remained the central thread of their ambitions. Their reassertion of this aim in 1614 provoked a complete breakdown of relations with the king. Sir Edwin Sandys reiterated the basis of their case, which was to establish freedom of trade as absolute property.

"Every man hath a propriety in his goods, which will not suffer that to be transferred to others without his own consent . . . [W]e have no propriety in our goods as long as the king's prerogative is to impose as much upon us as he pleaseth".[86] The challenge to the rights of the crown could not have been clearer. But the royal government had learned an important lesson in 1610, and the strategic error of allowing the matter to be aired was not repeated. In the Lords, the Bishop of Lincoln reverted to James's much sounder initial position and declared that it was seditious to question the royal prerogative in this way. The Commons confirmed the imperative nature of their own agenda by seeking to make supply conditional on the ending of impositions. The king responded by bringing the Addled Parliament to an end. There was an aftermath, which further illustrated the depth of the divide that was opening up between crown and Commons, between "patent" and "parliament". In the wake of the assembly, the king oversaw the burning of the written arguments that the Commons' leaders had intended to use to make their case against impositions in the Lords.[87] This act of destruction of the considered thoughts of the nation's representatives on their most vital concerns makes a revealing contrast to the care and commitment with which the merchants of Dartmouth had preserved exactly the same kind of record.

There appeared to be no way forward in confronting the king's rights with a simple assertion of their own claims. But it remained the central and consistent aim of the Commons to bring an end to impositions. It might have been supposed that in 1625, with the developing controversies in the field of foreign affairs, the issue of impositions would have slipped down the list of priorities, but its continued centrality was demonstrated when Sir Nathaniel Rich summed up the main conditions on which the House would be prepared to cooperate on the matter of supply. There were five principal stipulations, which contained not only the most current demands for a change in the course and leadership of foreign and military affairs but also the ongoing demand for a response to their grievance of impositions.[88]

They had taken the opportunity of the accession of Charles I to try a different and more subtle approach to prising away that prerogative. Charles expected, in this first parliament of his reign, to receive confirmation of the traditional grant of tonnage and poundage, which had become a customary right of the crown. By convention, parliament made this grant to the king for life at the beginning of each reign. This had been done for every monarch since Henry V and should have been a formality. In 1625, however, the Commons decided to exploit the theory of their grant and use it as leverage in their desire to undermine the prerogative of impositions. They determined that until they had received satisfaction in that regard, they would offer the grant of tonnage and poundage for one year only. Sir Robert Phelips flagged this up, proposing a saving clause be included in the bill of tonnage and poundage, to ensure "that nothing in this act may be passed against us for the maintenance of impositions".[89] The fact that Phelips and Coke were prepared to offer a retrospective endorsement of existing impositions if Charles gave up the right of setting them in future has sometimes been misconstrued as a non-confrontational attitude.[90] On the contrary, it would have met the Commons' essential aim of eliminating the prerogative of impositions and establishing the principle of representative consent in the customs dues. Thus when the matter of tonnage and poundage was discussed in a committee of the whole House on 5 July 1625, MPs linked their limitation of the grant directly to their desire to displace the right of prerogative customs generally. Some "alleged the pretermitted customs, grounded upon the misconstruction of that law . . . [O]thers to this added the question of impositions in general, and craved a special care not to have that excluded". Their solution in practical terms was that they themselves should assume responsibility for setting the book of rates. This was the logical conclusion of their ambitions and was encouraged by Sir Edward Coke. It could not, however, "be prepared in so short a time and sitting".[91] Moreover, the king showed no inclination to allow it, so the Commons held back their grant of tonnage and poundage.

It was really a kind of blackmail, but the Commons offered the inducement that the king's income from the customs might actually increase if they were given the power to set them. Thus, in 1626, Sir Nathaniel Rich assured Charles that "the subject would keep up his revenues according to the book of rates" if the king "would undertake that he be quieted from further impositions".[92] Otherwise the Commons could not see their way clear to confirming tonnage and poundage. They clung to this as their one real form of leverage, the only way that they could give some practical substance to their central purpose of bringing the customs dues under the control of representative consent. Not surprisingly, however, the king displayed little readiness to surrender his right of prerogative impositions in order to recover his right to tonnage and poundage.

In the parliament of 1628, Charles's dire need of money to finance his military policy obliged him to make at least a show of acceptance of the Petition of Right. In this, the Commons condemned as unlawful the king's recently exercised emergency powers of arbitrary imprisonment, forced billeting and discretionary taxation. The king responded only in the most vague and general terms that the laws and customs of the realm should be upheld.[93] But the Commons having, as they fondly supposed, established the principle that no money should be raised without parliamentary consent, now hoped that they could go on to settle that right in terms of the customs dues, their persistent radical ambition since 1610. In the wake of the agreement over the Petition in 1628, MPs reiterated their view that impositions were illegal and that the system of prerogative customs dues should be ended.[94] The proposed deal was the one that had been put forward by Commons' leaders like Sir Edward Coke and Sir Nathaniel Rich since 1625. They would confirm a grant of tonnage and poundage if the king would surrender his right of impositions and allow them to compile the book of rates by parliamentary authority. They asked the king for an adjournment so that they could consult their constituents about the appropriate levels.

Charles preferred to prorogue the assembly. He would certainly have liked his right in tonnage and poundage to be endorsed but not on the terms that were being offered. His consistent position on the matter was that he simply wanted the grant confirmed as it had been for his predecessors. There was no time allowed for MPs to establish the desired levels for the customs, and for this the king was held to blame. Sir Nathaniel Rich protested that "in this strait of time we have no way left . . . when if it had pleased the king to have adjourned we might have taken a different course . . . had his full revenue and the subject well contented".[95] Sir John Eliot expressed the same sense of frustration and resentment. As so often they were reduced to entrusting their demands to a remonstrance, declaring "tonnage and poundage and other impositions not granted by parliament . . . a breach of the fundamental liberties of the kingdom, and contrary to . . . the Petition of Right".[96]

Charles had the power and an urgent need to continue collecting tonnage and poundage notwithstanding the absence of parliamentary endorsement. In fact, he had closed the 1628 session with a speech stating his intention to do so as a right. Afterwards he called in the *Commons Journal* and unilaterally enrolled the speech, underlining his view of the matter. To Charles, tonnage and poundage was an undoubted right, which the Commons were seeking to deny him. So the 1629 parliament was called with the simple aim of getting the grant confirmed as it had been for his predecessors. The continued withholding of parliamentary approval had caused him some inconvenience, both practical and personal. Many merchants had responded to parliament's incitement to refuse the levy, in what has been called a "general strike". Furthermore, the king's reputation was suffering. The Tuscan ambassador was heard to observe "the sailors refuse to pay the usual duties, insisting that parliament did not grant them to the king as they did to his father".[97] Charles had powerful motives to try once more to get the monarch's traditional right to tonnage and poundage reinstated.

He prepared the ground by adding another gloss to his position, saying that he had only presumed to collect tonnage and poundage by necessity, not of right. This was an indication of what he proposed now, to give parliament another chance to make the customary grant. If the Commons would pass the bill as it had done for his predecessors, he would acknowledge that he raised tonnage and poundage by right of this grant, not by prerogative.[98] In other words, he simply wanted the levy regularized on the conventional basis. This took no account of what the Commons had been trying to achieve by withholding their approval since 1625. They gained nothing by the king denying that he had a prerogative right in tonnage and poundage. What they wanted was for him to deny his prerogative right of impositions.

So although some members were happy to set the tonnage and poundage bill in motion as the royal government wished, the dominant group in the House, led by Sir John Eliot, chose to defer it, while they focused on the new challenge to their liberties arising from recent controversies.[99] There was an obvious concern that the measure should not be driven through before the question of impositions had been dealt with. Secretary of State Sir John Coke was told precisely this when he tried to bring in the tonnage and poundage bill on 26 January.[100] Eliot noted that the bill in its present form left the power of impositions unchallenged. He also recalled that they might have resolved these matters in the previous session, if the king had allowed them the time.[101] This confirmed the condition on which they would have been happy to entertain the grant of tonnage and poundage—that is, if the king had been prepared to permit them to bring the customs dues in general under representative control. In the absence of this, the tonnage and poundage bill did not receive a reading. So if the proceedings of the 1629 assembly appeared somewhat unfocused, it was because the king and the Commons were completely at odds over this central issue, and neither could make progress with it. The Commons occupied their time with tangential matters of privilege and religion. To the king, these debates were "no more than a distasteful preliminary to a possible vote of tonnage and poundage".[102] Eventually, Charles concluded that such a vote would never take place and brought the session to an end. What the king had desired from the assembly was shown by his declaration following the dissolution, describing at some length how the grant of tonnage and poundage had always been taken as read by his predecessors.[103]

Thereafter, the king determined to rule without parliament. There was the general reason that the assembly had given him much more in the way of political aggravation than financial supply. The traditional relationship whereby parliament provided fiscal support for royal policy had broken down. But there was also the specific reason that he had failed, after a prolonged attempt, to induce the Commons to give him what he now wanted most—that is, the restoration of his rights in tonnage and poundage. And by the other side of that same coin, he had also failed to deflect them from what they wanted most—that is, the elimination of his right of impositions. There was little reason for Charles to call a parliament when it would only produce another round in that confrontation.

The assertion of a right of freedom of trade against arbitrary exactions and restraints was then the principal form in which the claim of absolute property was coming to supplant the medieval notion that property was conditional. An earlier chapter described the older view as summarized by Robert Crowley. He noted the essence of the traditional concept, which maintained that people were not free to do whatever they wished with their property, because they held it as a trust, with an obligation not to use it against the common good of their neighbours. In the long campaign against impositions, the Commons had asserted the opposite—that every man possessed complete propriety in his goods, against the king and all other men, to be used freely, and taken only by consent. Crowley had also noted the further proposition of conditional property—that all must give the king whatever he desired, for he was the sole judge of the public safety.[104] Crucially, however, as a concomitant of the assumption of absolute property and an inviolable right of consent, the king's exclusive role in determining the overall public interest was also beginning to be disputed.

NOTES

1. M. Overton, *The Agricultural Revolution in England*, (Cambridge 1996), pp. 22, 45, 133.
2. W.G. Hoskins, *The Age of Plunder*, (London 1976), p. 10.
3. A. Everitt, "The Open Market", in *Agrarian History of England and Wales 1500–1640*, ed. J. Thirsk (Cambridge 1967), pp. 490, 496, 500.
4. Ibid., p. 501.
5. Ibid., p. 499; C. Palliser, *The Age of Elizabeth*, (London 1983), p. 275.
6. B. C. Jones, "Westmoreland Packhorse Men in Southampton", trans. Cumberland and Westmoreland Antiquarian Society, LIX, (1960).
7. M. Overton, *The Agricultural Revolution in England*, p. 103.
8. *Statutes of the Realm*, IV, 2 & 3 Philip and Mary, cap. 13.
9. Daniel Defoe, *A Tour Through the Whole Island of Great Britain*, (Harmondsworth 1971), p. 496.
10. CSP Domestic, 1619–23, p. 130, 1631–3, p. 18.
11. C. Hill, *The Century of Revolution*, (London 1961), p. 20.
12. R.H. Tawney, *Religion and the Rise of Capitalism*, (Harmondsworth 1966), pp. 77–8.
13. J De Vries, *The Dutch Rural Economy in the Golden Age 1500–1700*, (Yale 1974), pp. 1–3.
14. F. Braudel, *The Mediterranean*, (London 1972), p. 213.
15. J. Chartres, *Internal Trade in England 1500–1700*, (London 1977), p. 13.
16. Ibid., p. 49; A. Everitt, "The Marketing of Agricultural Produce", in *The Agrarian History of England and Wales*, ed. J. Thirsk, IV, pp. 506–52.
17. M. Overton, *The Agricultural Revolution*, p. 133; J. Bohstedt, *The Politics of Provision*, (Farnham 2010), p. 68.
18. W. Harrison, "Of Fairs and Markets", in *Tudor Economic Documents*, ed. R.H. Tawney and E. Power, (London 1924), III, p. 73.
19. *A Discourse of the Commonweal*, (1549), ed. M. Dewar, (Virginia 1969), pp. 54, 57, 59, 90, 117–9.
20. *Piers Plowman's Exhortation*, (circa 1550), BL 1653/8, unpaginated but as counted, pp. 3, 7.
21. *Tudor Economic Documents*, ed. R. Tawney and E. Power, II, p. 188.
22. Ibid., p. 185.
23. Ibid., pp. 223–5.
24. Ibid.
25. J.E. Neale, *Elizabeth I and Her Parliaments*, (London 1954), II, p. 378.
26. *A Discourse of Corporations*, (1587–1589), in *Tudor Economic Documents*, III, pp. 266–7.
27. W. Notestein, *The House of Commons 1604–1610*, (New Haven 1971), p. 106.
28. Molin to the Doge, 23 June 1604, CSP Venetian, 1603–1607, pp. 161–3.
29. W. Notestein, *The House of Commons 1604–1610*, p. 108.
30. *CJ*, I, p. 218.
31. J. Kayll, *The Trades Increase*, (London 1615), BL 1138.b.10, p. 51.
32. Ibid., pp. 51–2.

33. John Chamberlain to Dudley Carleton, 8 July 1620, CSP Domestic, CXVI, p. 13.
34. E. Nicholas, *Proceedings and Debates in the House of Commons 1620–1*, (Oxford 1766), II, p. 303.
35. *CJ*, I, p. 539; *Commons Debates 1621*, ed. W. Notestein et al. (New Haven 1935), II, pp. 167–9; V, p. 272; VI, p. 31; E. Nicholas, *Proceedings and Debates 1620–1*, I, p. 122.
36. S.R. Gardiner, *History of England*, (London 1886), IV, p. 55.
37. C.H. McIlwain, *Constitutionalism, Ancient and Modern*, (Ithaca 1940), p. 138.
38. *Commons Debates 1621*, ed. W. Notestein, et al., IV, p. 272.
39. Ibid., V, p. 93.
40. *CJ*, I, pp. 520, 526, 537, 592, 620.
41. Ibid., p. 520.
42. Ibid., p. 592.
43. Ibid., pp. 526, 537.
44. Ibid., p. 529.
45. Ibid., p. 534.
46. C. Hill, *Reformation to Industrial Revolution*, (Harmondsworth 1969), p. 69.
47. *CJ*, I, p. 694.
48. Ibid., p. 591.
49. Ibid., p. 620.
50. *Commons Debates 1621*, ed. W. Notestein et al., III, p. 361.
51. Gerard de Malynes, *The Maintenance of Free Trade*, (London 1622), pp. 49–50.
52. Ibid., p. 51.
53. Ibid., pp. 52–4.
54. C. Palliser, *The Age of Elizabeth*, p. 192.
55. C. Hill, *The Century of Revolution*, (London 1961), p. 18.
56. R. Zaller, *The Parliament of 1621*, (Berkeley 1971), p. 130.
57. C.H. McIlwain, *Constitutionalism Ancient and Modern*, p. 138.
58. C. Russell, *Parliaments and English Politics*, p. 191.
59. Ibid., p. 196.
60. West Devon Record Office, Plymouth Corporation Archive, W48, f. 92.
61. R. Brenner, *Merchants and Revolution 1550–1663*, (Cambridge 1993), p. 206.
62. W. Notestein, *The House of Commons 1604–10*, p. 171.
63. *Tudor Economic Documents*, ed. R. Tawney and E. Power, II, pp. 223–5.
64. W. Notestein, *The House of Commons 1604–10*, p. 258.
65. Ibid., pp. 265–6; and S.R. Gardiner, *History of England*, II, p. 68.
66. *Proceedings in Parliament 1610*, ed. E.R. Foster, (New Haven 1966), II, pp. 100–6.
67. *Parliamentary Debates of 1610*, ed. S.R. Gardiner, Camden Society Old Series, lxxxii, p. 39.
68. Devon Record Office, Dartmouth Borough Archive, DD67721, f. 1.
69. W. Cobbett, *Complete Collection of State Trials*, (London 1809–28), II, pp. 483–6.
70. Ibid., p. 482.
71. Suggested, for instance, by J. Somerville, *Politics and Ideology*, (London 1986), pp. 51–3; B. Coward, *The Stuart Age*, (London 1994), p. 140; R. Brenner, *Merchants and Revolution*, p. 205.
72. Devon Record Office, Dartmouth Borough Archive, DD67721, f. 7; *Proceedings in Parliament 1610*, ed. E. Foster, II, p. 152.

73. Devon Record Office, Dartmouth Borough Archive, DD67721, f. 5.
74. Ibid. f. 7.
75. W. Cobbett, *Collection of State Trials*, II, p. 407.
76. *Proceedings in Parliament 1610*, ed. E. Foster, II, pp. 226–7; Dartmouth Borough Archive, Ms DD7723.
77. Thomas Hobbes, *Behemoth*, ed. F. Tonnies, (London 1889), p. 4.
78. For a detailed analysis of the record of the impositions debate preserved by Dartmouth Corporation, see G. Yerby, *People and Parliament*, (Basingstoke 2008), chap. 3 and app. 2.
79. An instructive but underused guide to the involvement of English landowners in trade is T.K. Rabb, *Enterprise and Empire: Merchant and Gentry Investment in the Expansion of England, 1575–1630*, (Cambridge, Mass. 1967).
80. D, Sacks, "The Corporate Town in the English State", *Past and Present 110* (1986), pp. 69–105.
81. M. Dobb, *Studies in the Development of Capitalism*, (London 1963).
82. R. Brenner, *Merchants and Revolution*, p. 205.
83. Ibid., p. 222.
84. Ibid. pp. 219–20.
85. C. Hill, *Intellectual Origins of the English Revolution*, (London 1972), p. 289.
86. *CJ*, I, pp. 472, 484.
87. S.R. Gardiner, *History of England*, II, p. 249.
88. *Proceedings in Parliament 1625*, ed. M Jansson and W. Bidwell, (New Haven 1987), pp. 414, 417–8, 419.
89. *CJ*, I, p. 813.
90. D, Hirst, *Authority and Conflict*, (London 1986), p. 140.
91. *Proceedings in Parliament 1625*, ed. Jansson and Bidwell, pp. 510–11.
92. "The Diary of Bulstrode Whitelocke 1626", 27 March, Cambridge University Lib. DD 12–22, f. 234.
93. *Constitutional Documents of the Puritan Revolution*, ed. S.R. Gardiner, pp. 66–70.
94. *Commons Debates 1628*, ed. R.C. Johnson et al., (New Haven 1977), III, p. 595.
95. Ibid., IV, pp. 449, 458.
96. Ibid., pp. 388–95.
97. R. Brenner, *Merchants and Revolution*, pp. 232–3.
98. J. Rushworth, *Historical Collections*, I, p. 642; *Commons Debates 1629*, ed. W. Notestein and F.H. Relf, (Minneapolis 1921), p. 11.
99. *Commons Debates 1629*, ed. Notestein and Relf, pp. 6, 113.
100. S.R. Gardiner, *History of England*, VI, pp. 34–5.
101. *Commons Debates 1629*, ed. Notestein and Relf, pp. 108–9.
102. C. Russell, *Parliaments and English Politics 1620–1629*, p. 412.
103. *Constitutional Documents of the Puritan Revolution*, ed. S.R. Gardiner, p. 83.
104. Robert Crowley, *The Select Works of Robert Crowley, The Early English Text Society*, extra series, XV, (London 1872), pp. 67, 132, 142–4, 157.

12 Parliament as a Point of Contact Between the Constituencies

The Emergence of a Freestanding National Interest, and the Roots of English Liberty

The readiness to deny the king a right to tax "upon any pretence of common safety" had an important further ramification, for if the king was no longer to be acknowledged as the exclusive judge of the overall public interest, then it followed that this important responsibility was being taken on more broadly, most obviously and significantly by those in the Commons and in the country who were insisting upon the right of consent.

The networks through which MPs and their constituents could arrive at an independently coordinated view of public affairs are very evident, though not always recognized as such. Modern historians have focused on the idea of parliaments as a "point of contact" between crown and people, an opportunity for monarch and subjects to come together, to make each other aware of their respective needs and concerns, and to offer mutual assistance. This is another of the ways in which revisionist academics have sought to minimize the potential for conflict between king and parliament, but the idea of harmony is obviously contradicted by the reality that from the beginning of the seventeenth century the assemblies were increasingly likely to result in confrontation.

This was largely because parliament had a strongly developing function as a "point of contact" between the constituencies themselves. The MPs' most natural and constructive area of contact was not with the king but with one another. The county and borough representatives were elected through an essentially uniform, kingdom-wide structure. Interrelations developed from county to county and from region to region, especially in the context of the rapid growth of internal trade during the sixteenth century. London served as a core connecting marketplace and the main outlet for the kingdom's first significant manufactured export, woollen cloth. It also became the base for political centralization, as both court and parliament, once itinerant, settled in the capital. London was unusual among European kingdoms, not only in its dominant position as the one major urban centre in the realm but also because it was both the capital city and the principal port. This encapsulates the unique basis for national integration that existed in England, in the combination of administrative and economic affairs. London was also the hub of social and cultural life and of legal business. So the sixteenth century witnessed a "landed classes' invasion of London", not least because "the profits and prestige of land made it a precious commodity to be fought over when the national land market was vastly increased by the sale of church and crown lands after 1540".[1]

At regular intervals, the gentlemen, lawyers and merchants who were the nation's representatives came together to formally compare notes. The House of Commons was a forum in which they could exchange views, confirm common interests, reach consensus on their requirements both general and particular and decide how they could best be provided for. This was indeed the process by which they made law, and from the 1530s onward it was a uniquely powerful form of law, often applied to the greatest national issues. It was not surprising that they came to feel that there was not much that was beyond their remit, and increasingly they found themselves identifying problems and solutions that were different from and sometimes opposite to those identified by the monarch. The potential for dissension was increased by the emergence of a series of national priorities in which MPs and their constituents could feel that they had a direct stake and which they could readily identify for themselves. Thus, over the course of the century there arose for the first time a set of freestanding or self-defining national interests—that is to say, public imperatives that could be perceived and established independently, without necessary reference to the preferences of the crown.

The existence of such vital concerns is apparent from the 1530s onward. The triumph of parliamentary law had destroyed the independent jurisdiction of the medieval church, and produced a conscious declaration of national independence. In the process, it had also generated a number of autonomous public interests and some powerful national prejudices. Although it did not necessarily involve a theological divorce, it immediately put in place the antipapal xenophobia that became a constant obsession of the English people for centuries. This is an example of a self-defining national interest of a type that could not have existed before. Now it took shape as a public instinct that could be felt separately from the interests of the crown. In fact, popular opinion proved to be more aggressively anti-papal than each successive royal regime, to a greater or lesser extent, and this was regularly asserted in ways that constituted a contradiction of government policy.

It was of some importance that although anti-papal feeling had no immediate theological basis, it did serve the economic self-interest of English landowners. The Papacy was hated not simply because the displaced Roman Church and its allies were a self-evident threat to the new nation state and national church but also because the restoration of papal authority would put the laity's tenure of the church lands at risk. Nothing consolidated the idea of a unified, sovereign "land" so much as the luck of having just acquired a goodly portion of it, along with the fear that its dispossessed owners across the sea would take it back again if they could.

So the maintenance of the freedom of the new nation state against external threats immediately came to incorporate other more specific notions of liberty, like the freedom of property, which in turn implied a new emphasis on specific political liberties, like rights of consent. Furthermore, although the break with Rome had not involved an immediate theological breach; it had created a platform for what the Reformist movement called "Gospel freedom", stemming from the predominantly evangelical character that the English Reformation had always displayed and indicating the need for a more open relationship with the Bible.

These perceptions were consolidated into a coordinated national outlook partly because they could all be defined in antithesis to one particular external threat—the power of Spain. The peculiar strength of English antipathy to Spain first became evident at the beginning of the reign of Mary Tudor, in the outspoken opposition to her proposed Spanish marriage. Historians have sometimes expressed surprise that this public animosity was so firmly in place even before the events of Mary's reign unfolded. David Loades wondered how that sentiment developed, since there had been little direct contact with Spain in recent times.[2] But it is not really hard to understand. In one respect, it was a simple process of deduction. During the 1530s and 1540s, Spain emerged as the dominant power in Europe, the champion of the old church, the principal opponent of Gospel freedom and the one power with the resources, military strength and singleness of purpose to win back lands, in every sense of the word, for the Pope. Spain thus became the specific threat to all the characteristic rights and interests that were now being assumed by the English. For good measure, Spain was the main obstacle to the freedom of the seas. In the middle of the century, the English people were launching themselves towards a perceived and potentially glittering future in oceanic trade and colonization. The blot on the horizon was the imperial power of Spain. Opposition to Spain became the defining feature of the new English nation's outlook on the world.

This logic was reinforced by some real differences of principle and political practice. The Spanish ambassador, unable to ignore the general hostility that he and his fellows encountered at the beginning of Mary's reign, put it down to the picture of the Spanish as being cruel and tyrannical that English traders received from their German and Dutch contacts and "the manner in which your majesty's own subjects complain of their arrogance".[3] This was based on more than general impressions. It was understood that the Spanish system of government contradicted at every point the preferred English alternative, and there was substance in the contrast. Whereas in England it was now accepted that sovereign law could only be made in parliament, in the core Spanish kingdom of Castile, the monarch was still the sole lawmaker. Similarly, the Spanish crown could in effect raise taxes without the consent of the Castilian Cortes. Furthermore, the taxes fell most heavily on the poor, again the opposite of the situation in England.[4] The poll tax had played a part in provoking the Peasants' Revolt of 1381, and although the rebellion was suppressed, it was a shock to the establishment, and no further attempt was made to shift the burden of taxation onto the shoulders of the poor. The proportion of people paying tax in England continued to diminish during the sixteenth century, and there were some places where only one in every dozen households paid direct taxation. Those higher up the socio-economic scale, who paid the subsidies, had control of setting the levels. Francis Bacon observed that "the Englishman is most master of his own valuation, and the least bitten in purse of any nation in Europe".[5] Most crucially, these arrangements gave the gentry a direct interest in maintaining and extending the principle of consent and encouraged them to step forward as the champions of that developing aspect of English liberty.

Spain presented the most powerful and comprehensive contrast, but not the only one. English liberty revolved around the idea that the kingdom was peculiarly fortunate in these respects. Thus it was assumed that other European kingdoms were generally subject to money-raising practices of which England was relatively free and would wish to be entirely so. This was the view given voice by John Hales, advising the government against a new imposition on the cloth trade in 1559. "I believe there is no Englishman that knoweth the government of other countries and the great impositions wherewith the people are yoked, but perceiveth and confesseth that this realm in respect of others is only free, and will desire that it may so continue".[6] Again there was a real element of truth in this. As already noted, England was unusual in that "the great bulk of buying, selling and transporting of goods was free of all imposts".[7] At first, this was just a fortuitous circumstance, highlighted by the rise of the cloth trade. But it became a crucial aspect of English liberty to consolidate and extend that freedom as a settled right.

The nature of "English liberty" has been somewhat obscured by the mistaken tendency of some Whig historians to backdate it to the feudal constitutionalism of the medieval period.[8] The notion of liberty most distinctively associated with the English is a rather more bourgeois concept. It had its roots in the mid-sixteenth century and was advanced and contended for in the decades before the Civil War of the 1640s. It hinged in various ways on the elevation of rights of consent. A crucial coordinating aspect was the establishment in the 1530s of a fully sovereign, omni-competent statute law, which drew its binding force from the capacity of parliament to reflect the consent of the entire realm. The doctrine that statute law was binding upon all because all had consented became current in legal circles towards the end of the fifteenth century, as the Commons acquired an executive role in legislation. This authority was then employed by Thomas Cromwell to destroy the independent legislative power of the church, on the basis that only parliament, through its representative function, could validly bind people in law. Thus, the representation of "everyman" was established as the authorizing principle of the one, indisputable sovereign law. In the decades ahead, the idea that the exercise of consent was necessary for the "absolute" power of legislation came to structure parliamentary ambitions. Previous chapters have described how the parliamentary attack on the prerogative of impositions in 1610 was predicated on the supposition that all such "absolute" powers should be subject to representative consent, and the first aim of the revolution of the 1640s would be to give parliament an automatic position in the polity, commensurate with its defining place in legislative sovereignty.

Developing in unison with the growing political status of rights of consent was the most characteristic material foundation of English liberty—the desire to establish a right of freedom of trade, which became the principal, practical motivation for enhancing the authority of parliament. The ambition derived from the exceptional degree of actual trading freedom that existed in the sixteenth century, and we have seen it emerging as a principle from the late 1540s. The demand to be left free of arbitrary exactions became central to the drive to make property "absolute", and it was asserted with increasing controversy against a royal government that remained wedded to the concept of "regulating" trade by impositions and monopolies. Land taxes had long been subject to a right of consent, in theory, but this was quite nominal until the political nation began to identify its own definite areas of preferred policy in the second half of the sixteenth century. It was a further step in establishing a right of *real* consent when the Commons began, in 1610, to challenge the residual power of the crown to raise money by prerogative in the case of public safety.

The other important feature of English liberty was Gospel freedom. This was manifested in part by the tolerationist trend already noted. In the light of this, the characterization of the Spanish as "cruel" was not a vague generalization, but a specific critique. These distinctive tendencies in England have not received due recognition from modern historians, who tend towards the restrictive view that the typical sixteenth-century provision of the single-faith polity precluded the emergence of genuine tolerationist sentiment.[9] This obscures the important psychological changes that were developing beneath the surface. It is clear that there was a tolerationist inclination taking shape, based on the principle that faith could only come by consent to the Gospel. The tolerationist trend must be recognized not only for its own sake but also because it reflects other aspects of the transitions that are under discussion in this study. It illustrates the way that the medieval context, based on the undifferentiated continuum of all things, was losing ground to the view that there was a clearer distance and distinction to be placed between the physical and intellectual spheres. This highlights the common feature of the positive force of English liberty, which embraced the principle of consent as the antidote to compulsion in every field.

The gradual emergence of a tolerationist view is perhaps the clearest demonstration of the elements of a right of consent taking shape. In 1530, the Commons' remarkably assertive attack on the arbitrary nature of heresy proceedings was the first indication that the power of the hierarchy to simply impose beliefs was being challenged in the name of private judgement. The demand for due process and the right of the accused to make a case contradicted the basic assumption of inquisition, which was that the individual mind did not possess an independent integrity and that the clergy, as instruments of the God-given truth, were free to enforce their own interpretation. In defiance of this, the Commons' petition declared that the accused should not be destroyed simply for expressing their own "innocent truth", which carried the radical implication that even though this "innocent truth" might not suit those in command of "the high mysteries of the faith", it might nevertheless be valid.[10]

The essential principle that truth could be reached only by individual assent to the Bible was enunciated by Henry Brinkelow in 1542, in his protest against the burnings of Robert Barnes, William Jerome and Thomas Garret. These persecutions displayed all the unacceptable, arbitrary features of heresy proceedings that the Commons had begun to expose a decade earlier, and Brinkelow reiterated the demand for due process. He went on to propose that these matters should be dealt with not by inquisition, but as a debate between equals, "tried by an indifferent judge". This defied the assumption of the clerics that they were necessarily correct and could impose their view. The core tolerationist proposition that consciences could not be forced was proclaimed in Brinkelow's insistence that physical punishment could be applied only for temporal transgressions "but not for his faith's sake, for faith is the gift of God only".[11] The power of compulsion was being ejected from the sphere of conscience.

In the mid-sixteenth century, in *A Discourse of the Commonweal* and most clearly in the work of William Turner, came the first explicit definition of the mind or conscience as an independent, inviolable sphere that neither should nor could be subject to physical determinants. "Saying is no material thing that we must fight withal". Heresy must be combated with spiritual, not physical fire. "It were most meet that we should fight with the sword of God's Word, or else we are likely to prosper but a little in our business".[12] This was the essence of the change. The special stress that the English Reformation placed on rejecting externalist forms and embracing an evangelist Gospel freedom had led to the belief that the mind was an autonomous sphere and that consciences *could* not be forced. Faith must come by consent. Turner boldly went on to outline the broad assumption of liberty that this involved, taking it into a political dimension. He was prepared to assert that the individual was not obliged to follow whatever the authorities prescribed, but only what he himself could accept as recommended in the Bible.[13]

Turner, Brinkelow and *A Discourse* all espoused the proposition that disputes about the truth could not be settled by force but might be settled by argument and debate. In Elizabethan times, Richard Hooker gave authoritative endorsement to this idea and outlined the specific procedure of rational demonstration on which it must rest. The individual was not required to "yield unto anything other assent than such as doth answer the evidence". Hooker also showed how the process of individual consent underpinned the concept of sovereign representative law. Parliamentary statute was binding on all because all had consented to it. "Although we be not personally ourselves present, notwithstanding our assent is by reason of others agents there in our behalf".[14] Thus the principle of consent bound the elements of English liberty together.

The changing balance of initiative between the intellectual and the physical spheres and between the individual and the hierarchy found central expression in the great Repeal Act of 1547, which boldly removed the heresy laws from the statute book and liberalized Henry's draconian treason legislation. "Unquestionably, these measures were, and were intended to be, a manifesto of freedom".[15] But what was the essence of this freedom? At the passing of the act, Protector Somerset told the imperial ambassador that some of the laws as they stood "are almost iniquitous in their severity".[16] The preamble of the act suggested that the new reign had brought a less troubled climate in which such repression was no longer required. This seemed a somewhat naïve assumption, when the challenges to the new order were probably increasing. Some historians have thought indeed that, "a wholesale abrogation of such weapons . . . was idealist folly".[17] This underestimates the stabilizing effect of the support that the measure commanded within the realm. But the willingness to relinquish the most powerful tools of control certainly underlined the element of idealism. The happier climate had more to do with an emerging balance of common principle than with a sense of settlement or security.

The Repeal Act was a combined effort to which the Protector and the parliament were equally committed. It seems that Somerset took the lead in the general initiative to remove Henry's most repressive laws and was personally responsible for the clause that changed the law so that words could not be construed as treason on the first offence but only after a third instance.[18] Some of the other clauses appear to have emerged from consultations with the Commons. These included the very striking provisions that henceforth confessions must not be induced by force. "No person or persons . . . shall be indicted, arraigned, condemned or convicted for any offence of treason, petty treason, misprision of treason, or for any words before specified to be spoken . . . unless the same offender, speaker, offenders or speakers be accused of two sufficient and lawful witnesses, or shall willingly without violence confess the same".[19] This injected a vigorous burden of proof where none had previously existed. Treason trials were normally held under a specially appointed commission, which made it easy to rig the court. The accused had no proper rights of defence, one witness was taken as proof of guilt, and torture was routine. It becomes possible then to suggest wherein the essence of this "manifesto of freedom" lay. The Repeal Act introduced the principle of due process to an area traditionally governed by arbitrary force. It shared the common assumption of liberty that informed the parliamentary attack on the arbitrary nature of heresy proceedings in the early 1530s and echoed thereafter through the work of Henry Brinkelow and William Turner. The practice of compulsion was being questioned in every field.

It was no accident that the Repeal Act also saw the end of the heresy legislation, Henrician and Lancastrian. Religious persecution was no longer to be endorsed by the sovereign law of England. This appeared indeed to be a distinctive English liberty, running counter to the state of affairs in other European kingdoms. The government was consciously surrendering traditional powers that were elsewhere still being exercised to the full. The Emperor Charles V was currently embarking on another round of persecutions and burnings in his German lands, to accompany the similar campaign that he had been pursuing in the Netherlands. In France too the flames of persecution were rising highest at this time. On the Protestant side, John Calvin was busy making Geneva a byword for intrusive intolerance.

The situation in England was different, and historians were once happy to acknowledge it. Thus W.K. Jordan believed that under the Protectorate, "England was to undertake a precocious trial of a policy of unofficially announced religious toleration". The repeal act displayed "the spirit of this remarkable and magnanimous man".[20] More recent historians, having burdened themselves with the illogical proposition that the past cannot prefigure the future, were driven to attack Somerset's advanced reputation.[21] M.L. Bush, who began the assault, openly admitted this destructive intent, apparently convinced that anything that erased the signs of change was justified. His very restrictive interpretation presented the unique absence of persecution in Somerset's reign as just an unintended accident.[22] This was an unfortunate judgement on a man who should probably be credited with the first tolerationist policy in English history. It illustrates the damage done when historians set out to disprove the idea that the past can lead to the present.

There can be little doubt that Somerset's policy reflected a conscious tolerationist stance. He is known to have recommended tolerance to other kingdoms, though few of them followed the advice. In England itself, the Protectorate was the only regime in the sixteenth century that was effectively without persecution. No one died for their beliefs, and a series of dissenters who would have been burned in the previous reign were pardoned during Somerset's rule. Moreover, Catholic activists were treated as leniently as Protestant dissenters. As Jordan says, they were allowed to publish freely: "[N]one were done to death, very few were imprisoned, even for a short season".[23] Even a group of Anabaptists escaped serious punishment. Bush suggested that this was merely because Somerset fell from power before he could get around to burning them. Fact and logic make it much more likely that he was following the advice of his chaplain, William Turner, that ideas were "no material thing that we can fight withal". Somerset sought, specifically, to achieve reconciliation through the reception of scripture. To facilitate this, Henry's restrictions on Bible access for the lower classes were removed. It was perhaps naïve to think that opening the Bible to all would produce a generally agreed interpretation. Gospel freedom would often have to be compromised in the cause of public order. But the crucial principle was established that faith could come only through free consent to scripture.

Jordan identified the changing psychology behind this. The Duke, he suggested, did not believe that "force was a proper or useful instrument of religious policy".[24] The assertion of a right of consent against compulsion or imposition was indeed the common and coordinating feature of the various aspects of English liberty as it began to arise in the mid-sixteenth century and came to be asserted by revolution in the middle of the seventeenth. This has not received adequate treatment from modern historians. Sometimes the emergence of English liberty is implied but not defined. Thus David Loades noted that the burnings of Mary's reign enabled "skilful propagandists to cast the blame for such unprecedented severity on the unpopular Spaniards, and to represent the Protestants as martyrs to the cause of English liberty as well as to their faith".[25] But in what did English liberty actually consist? What made it a relevant and meaningful concept?

The tendency to leave the matter vague leads to the mistake that it was about conserving ancient liberties. Thus a recent attempt to identify the popular freedoms involved in the English Revolution fails to see through the inevitable rhetoric of precedent and reaches the unsatisfactory conclusion that there was a general idea of the freeborn people "embedded in the myth of the Ancient Constitution".[26] The notion of the freeborn people was important as propaganda, and Christopher Hill described the wide resonance that it had in the 1640s. But if we allow ourselves to conclude that a myth of ancient freedom was the liberty being fought for, we are in danger of failing to recognize the real and truly radical liberties, like freedom of trade, rights of consent and liberty of conscience, that were actually at issue.

In fact, there does exist a clear definition, from the time of the Civil War itself, when the liberties of the English were proved in action. The analysis came from Gerrard Winstanley in the wake of the conflict, in his tract *The Law of Freedom*. Everyone was talking about freedom, he said, but what did they actually mean by it? Winstanley is an unusually detached voice, for his own favoured concept of liberty was different from most, being the rather un-English notion of the common ownership of the land. From this moral high ground, he observed the other kinds of liberty for which most of his fellow parliamentarians had been striving. The essential political platform was the installation of "successive parliaments" in order to ensure the consistent exercise of supervisory consent over the processes of government. This principle, embodied in the Triennial Act, was the core proposition of radicals like Winstanley and the Levellers, just as it was of mainstream parliamentarian leaders, from John Pym to Henry Ireton. Winstanley also had a clear view of the particular, practical liberties sought by the parliamentary cause through the establishment of rights of consent, and they were threefold. First in line was freedom of trade—"the free use of trading and to have all patents, licenses and restraints removed". Associated with this was the assumed liberty of the gentry and the freeholders to be "landlords of the earth" with exclusive rights over their property. The other freedom was liberty of conscience, for people to access the Gospel for themselves, without being compelled to any particular form of worship.[27] These were also the distinctive set of liberties that can be seen appearing in the middle of the sixteenth century.

Throughout the hundred years between the emergence of English liberty in the mid-sixteenth century and its trial by battle in the mid-seventeenth, its defining antithesis was the power of Spain. When, at the end of 1553, Mary's intention of marrying Philip of Spain became clear, a parliamentary delegation attempted to dissuade her. The queen angrily and personally rebuked them for meddling in matters that she regarded as her affair alone. She was well within her rights. Traditionally, relations with other kingdoms had been the preserve of the monarch, and foreign policy had been determined by the personal and dynastic preferences of the crown. What was new was the emergence of an independently definable national interest and, in consequence, a developing public desire to assert it.

The parliamentary approach having failed, some parties turned to more physical means of expression. In the New Year, the street boys of London greeted the arrival of the imperial embassy by pelting them with snowballs.[28] More powerful figures had more solid projectiles to hand. The uprising known as Wyatt's Rebellion began on 25 January 1554. The name makes it sound like a personal crusade, but in fact it was of a strongly national complexion. It should have been accompanied by three simultaneous risings in the West Country, the Midlands and East Anglia, led by prominent figures of the Edwardian Privy Council, who were reluctant to see the distinctive public policies of those years reversed.

In the event, Wyatt was obliged to act prematurely and in isolation. But his propaganda reflected the national frame of reference. He asserted the concepts of "commonwealth" and "liberty" that had become current in the Edwardian years and declared his aim to be "the avoidance of strangers". A Norwich carpenter joined Wyatt in the belief that, otherwise, "the Spaniards should have our houses, and we should live like slaves".[29] Despite being the lone flag bearer, Wyatt might well have succeeded. There was a good deal of support in London, and he might well have been able to enter the city if he had acted more decisively before Mary's supporters had gathered. Even in defeat the national perspective of his initiative did not go unacknowledged. When he was executed, on 11 April 1554, people dipped their handkerchiefs in his blood, treating him as a martyr.[30]

It has been recognized that, "although the rebellion failed, it marks a turning point in English politics. Previous Tudor revolts had sought a remedy for grievances, not power. Wyatt and his followers intended an alteration of the monarchy, and they fought not merely for dynastic purposes, but for issues".[31] This was indeed a crucial development, but, as so often, the full implications have not been acknowledged or followed up by historians. What the rebellion actually represented was the first assertion of an independent view of the national interest. This perspective was extended and consolidated in the decades to come, with profound consequences for the future of the English polity. It is a central task of the present study to trace that process.

The events of Mary's reign and the consequences of the Spanish alliance that they were thought to reflect did much to reinforce the preferences and fears of the political nation. The queen experienced the least difficulty in getting the liturgical and theological aspects of the old church restored. Although there was significant opposition—from more than a quarter of the House of Commons—nevertheless her first parliament agreed to repeal the Edwardian Act of Uniformity, and the official beliefs of the kingdom returned to what they had been under Henry VIII. But that was as far as parliament would willingly go. A series of bills to revive the heresy laws were either defeated or deflected, much to the anger of the queen. And any hint of a programme for returning the abbey lands was dismissed out of hand. The parliament that met on 5 October 1553 made it clear to Mary that she was not at liberty to divest herself of the supremacy or to restore papal authority and that the church lands must remain with their new owners. Renard perceptively observed to his imperial master that there was an underlying problem here: "[T]he Catholics hold more church property than do the heretics".[32]

Drastic measures were required to overcome the problem. Steps were taken to ensure that the next parliament, which was to meet on 12 November 1554, would be more amenable. By then, the Catholic bishops were back in place, and there had been an attempt, for the only time in the Tudor century, to "pack" the House of Commons with government supporters. Even so, the most substantial of concessions was required for the queen to get her way. Parliament would agree to revive the heresy laws and the papal supremacy only if the gentry's tenure of the abbey lands and impropriated livings was guaranteed. So, with the blessing of the papacy, the laity's hold on the properties of the old church was put under the powerful protection of English statute law, and Mary was given the same level of authority for a campaign of persecution. It was a bargain of which the devil himself might have been proud. The gentry established an unshakeable grip on their plunder, and Mary recovered her right to burn people at will.

Oddly enough, Mary was perhaps the least to blame, for the rules of her church still told her that purification by fire was a way to preserve the truth. The parliamentary gentry had allowed their love of land to override their dislike of persecution, though they may still have hoped that Mary would moderate her actions in line with her promise at the beginning of the reign that consciences would not be forced. When it became clear in 1555 that the queen intended to use the heresy legislation to persecute and burn dissenters without restraint, parliament again launched a vigorous protest. The MPs were reflecting the general sentiment of their constituents. Public reaction to the burnings was one of clear disapprobation and distaste, confirming that Somerset's experiment in toleration had been underpinned by a basis of public approval. Mary's programme of persecution was generally thought, with understandable logic, to reflect the influence of Spain. In fact, the imperial ambassadors Renard and Michaeli were so alarmed by the public reaction and by the impending public relations disaster that they pleaded with Philip to restrain the zeal of the bishops and the queen. Philip seems not to have heeded the advice.[33] The association between the Spaniards and the exercise of cruel, arbitrary force was in consequence graphically underlined.

Public opposition was apparent in all ranks of society, and, like the defence of the appropriated church lands, it was not essentially Protestant in outlook.

> Some of the recorded instances of distaste and distress derive from people who did not share the religious views of the victims, while ample evidence remains that sheriffs, justices and gentry were reluctant to carry out these acts or grace them with their presence. In the latter stages, demonstrations of sympathy by the crowds towards the sufferers became so frequent that a proclamation forbade them on pain of death.[34]

That Mary ended by threatening the general run of her subjects with death for disapproving of the burning of others was proof enough that she was contradicting a powerful trend of opinion within her kingdom.

One particular aspect of the protests clearly illustrated the principled objection that had arisen to the process of inquisition. "The role of the clergy in the proceedings was particularly resented . . . there was indignation at the way the ecclesiastical authorities conducted their enquiries in the church courts, delving into the consciences of those who were brought before them". This carried striking echoes of the Commons' original attack on heresy proceedings in the early 1530s, deploring the fact that the clergy were able to trap people by "subtle interrogatories". In essence, it was no longer acceptable that the clergy should deny the individual mind its own integrity and impose their will upon it. The mind or conscience was being redefined as a private sphere, with rights of consent. On this basis, "unlike contemporary Spaniards, English people no longer regarded heresy as a terrible crime".[35] On the contrary, they were coming to feel that the persecution was the crime. Thus, another dominant English phobia, against the very concept of inquisitions, was taking shape.

The rejection of arbitrary force was, then, the common element of the various aspects of a developing concept of English liberty. In every sphere of activity, there was the growing assumption of an inviolable self-propriety. In religious affairs, there was a broadening belief that consciences should not and could not be forced. In economics, as the positive side of agrarian individualism, there was a growing determination that property should never be taken by force—there must always be specific consent. The same principle lay behind the demand for a right of freedom of trade against impositions. And so it was in the field of lawmaking, for the one way that law could be regarded as applied without force was when it was made through a system of elected representatives entrusted with the consent of their constituents. English liberty was not just a reaction against certain arbitrary practices. It embraced a positive, alternative principle. In all its aspects, it revolved around the assertion of a right of consent.

So in the reign of Mary Tudor, the English political nation had a close encounter with everything that it was coming to consider alien to its interests and inclinations. The Spanish marriage, the campaign of persecution, the threat to bring back the papacy and take back the church lands, and the French war in which the English navy kept the channel safe for Spanish forces, all contradicted the most notable trends of public opinion in England. Ironically, the experience of the reign served to confirm the political nation in its own distinctive character and in the belief that to preserve and further it, the power and influence of Spain had to be resisted. And Mary's reign had another important lesson to teach, closer to home. It was a comprehensive demonstration that the preferences and policies of the crown could be specifically at odds with the perceived interests of its subjects. It should be said again that this was not in some sense Mary's fault. Traditionally, the kingdom's relations with the outside world had been determined and defined by the personal and dynastic preferences of the crown. Mary was arguably less to blame than anyone for creating the circumstances in which, since the 1530s, the national interest had taken to defining itself. But by that very token she was peculiarly fated to reinforce it by contradiction and to confirm its existence as an independent perspective.

The divergence between Mary and her people was also reflected in the structures of government. The question of the scope and scale of Mary's council has been "nuanced" this way and that by historians until little in the way of a clear outline remains. But certain points of real significance can be suggested. The development of the Privy Council in sixteenth-century England was an important change in political organization, only marginally exaggerated by Geoffrey Elton as the core of a revolution in government. The ungainly, amorphous king's council that had survived into the reign of Henry VII was transformed during his son's reign into a streamlined, expert body with an independent status of its own, capable of acting as a coordinated national control centre. The new arrangement was in place by the 1530s. "Cromwell observed the birth of the elite board of office holders, which by Elizabeth's reign had assumed corporate responsibility for Tudor government under the crown and limited the role of the household in favour of national administration".[36] In the reign of Mary, this system was allowed to lapse. As many historians have noted, she relied on a small circle of devotees, and deprived herself of the broader, more objective advice of the Privy Council. This necessarily prejudiced the balance of her policies, though more recently it has been suggested that the fact that Mary and her consort Philip of Spain made policy without reference to the Privy Council was not a sign of weakness or inadequacy and that in any case their court-based inner circle included many of the experts who had served on Edward's councils.

It is true that Mary and Philip, as personal monarchs, had a perfect right to make policy within the confines of their court, and there was no intrinsic reason why this should be inadequate. But sidelining the Privy Council had particularly negative consequences in the English context. An important development in recent historiography has been to bring the role of Philip into clearer focus, and one of the innovations made by the Spanish consort was to formalize the inner court circle into a Select Council. This, as Glyn Redworth has pointed out, reflected the continental model of government.[37] It was an exclusive body, subsumed in the court, which coordinated policy across other departments of state. The Privy Council in England, as it emerged in the time of Thomas Cromwell and took definitive form in the reign of Elizabeth, had a significantly different role. It was not just the main arm of government but also the eyes and ears. It was the first level of connection between the court and the country, interacting with authority and society on the ground. Prominent local landowners were represented in council, and Privy Councillors were placed in offices and commissions in the counties. "Cromwell's policy centred on the coordinated use of nobles, assize judges, privy councillors, courtiers, bishops, JPs, and mayors and aldermen of towns".[38] The Privy Council in this form was a means of aligning policy with the body of the kingdom at large. By discounting this facility, Mary put a further, structural distance between herself and her people.

NOTES

1. *London 1500–1700*, ed. A.L. Beier and R. Finlay, (London 1986), pp. 12–3.
2. D. Loades, *The Reign of Mary Tudor*, (London 1979), p. 70.
3. Renard to the Emperor, 11 December 1553, CSP Spanish. *XI*, p. 425.
4. Henry Kamen, *Spain 1469–1714*, (London 1985), p. 14.
5. C. Palliser, *The Age of Elizabeth*, (London 1983), p. 12.
6. John Hales to William Cecil, 20 March 1559, in *Tudor Economic Documents*, ed. R. Tawney and E. Power, (London 1924), II, pp. 223–5.
7. C. Palliser, *The Age of Elizabeth*, p. 12.
8. The underlying supposition of G.M. Trevelyan, *England Under the Stuarts*, (London 1945).
9. For instance, Ole Peter Grell, *Tolerance and Intolerance in the European Reformation*, (Cambridge 1996), p. 6.
10. Petition of the Commons, 1531, TNA, SP6/7, ff. 220–1.
11. Henry Brinkelow, *The Complaint of Roderick Mars*, (1542), EETS, e series 22, pp. 29–32.

12. William Turner, *A Preservative or triacle against the poison of Pelagius*, (London 1551), p. 4.
13. William Turner, *A New Dialogue*, (London 1548), pp. 13, 60.
14. Richard Hooker, *Of the Law of Ecclesiastical Polity*, ed. C. Morris, (London 1909), vol. II, bk. 5, pt. 10, p. 37; vol. I, bk. 1, pt. 10, p. 194; vol. I, bk. 2, pt. 7, p. 268.
15. G.R. Elton, *England Under the Tudors*, (London 1967), p. 205.
16. CSP Spanish, 1547–1549, p. 205.
17. G.R. Elton, *England Under the Tudors*, p. 205.
18. M.L. Bush, *The Government of Protector Somerset*, (London 1975), p. 135.
19. *Statutes of the Realm*, I Ed VI, cap. 12.
20. W.K. Jordan, *Edward VI: The Young King*, (London 1968), pp. 127, 172.
21. M.L. Bush, *The Government of Protector Somerset*, (London 1975): D. E. Hoak, *The King's Council in the Reign of Edward VI*, (Cambridge 1976); J. Guy, *Tudor England*, (Oxford 1988), pp. 201–8.
22. M.L. Bush, *The Government of Protector Somerset*, p. 118.
23. W.K. Jordan, *Edward VI: The Young King*, p. 219.
24. Ibid., pp. 126–7.
25. D. Loades, *Reign of Mary Tudor*, pp. 275–7.
26. D. Underdown, *A Freeborn People: Politics and the Nation in 17th Century England*, (Oxford 1996), pp. 131–2.
27. *Winstanley: The Law of Freedom*, ed. C. Hill, (Cambridge 1983), pp. 288, 294.
28. J.D. Mackie, *The Earlier Tudors*, (Oxford 1966), p. 537.
29. *Depositions Taken Before the Mayor and Aldermen of Norwich*, ed. W. Rye, (Norwich 1909), p. 56.
30. J. Guy, *Tudor England*, p. 232; D. Loades, *Mary Tudor*, pp. 95–6, 124–9 and *Two Tudor Conspiracies*, (Cambridge 1965).
31. P. Williams, *The Later Tudors*, (Oxford 1995), p. 96.
32. Renard to the Emperor, 3 September, cited in J.D. Mackie, *The Earlier Tudors*, p. 544.
33. D. Loades, *The Reign of Mary Tudor*, pp. 175–6.
34. A.G. Dickens, *The English Reformation*, (London 1964) p. 268.
35. A. Somerset, *Elizabeth I*, (London 1991), p. 48: D. Loades, *The Reign of Mary Tudor*, pp. 275–7.
36. J. Guy, *Tudor England*, p. 163.
37. G. Redworth, "Matters Impertinent to Women: Male and Female Monarchy Under Philip and Mary", *English History Review 112* (1997), p. 601.
38. J. Guy, *Tudor England*, p. 168.

13 The Elizabethan Nation

"The Envy of Less Happier Lands"

In Mary's reign, the emerging, dominant interests of the political nation had been reinforced by contradiction. In every aspect of her policies, she had endeavoured to bring the ways of Spain to England. The two Houses of Parliament had to make firm common cause to prevent the installation of Philip as her successor in place of Elizabeth. Ironically, Mary had merely succeeded in crystallizing the general antipathy to Spain. The only thing that the English people wanted to imbibe from the Spanish was their empire. Under Elizabeth, these popular preferences and prejudices were consolidated more positively, but also in a way that underlined their autonomous power. During the course of her long reign, the political nation found itself in the kind of relationship with Spain that they were coming to perceive as natural and necessary—that is, a state of conflict. Furthermore, they managed to meet the challenge with some success, which they took as vindication.

That is not to say that Elizabeth immediately embarked on a policy of confrontation with Spanish power. Even had she wished to do so, caution was her natural mode and was an approach well suited to the current needs of her kingdom. But in her first exchanges with parliament, the queen was noticeably anxious to provide assurances that her greatest study would be the wishes of her people and that there would be no repetition of the hated and disastrous Spanish marriage.[1] At the opening of her first parliament, her representative the Lord Keeper drew a contrast in "guarded but unmistakable words" between the unfortunate policies of the late queen and such a princess as they now had "to whom, nothing—what nothing?—no, no worldly thing under the sun is so dear as the hearty love and goodwill of her nobles and subjects".[2] The message was clear: There was a perception that Mary had placed herself in some degree of opposition to the inclinations and interests of her people, and Elizabeth would not make the same mistake.

The change of direction to some extent dictated itself. Not the least ironic effect of Mary's policies was that they virtually obliged Elizabeth to do the opposite. The enduring memories of Mary's reign were of disease, bad harvests, high taxes, violent religious conflict and a disastrous foreign policy. This had left much to be desired. So Elizabeth's frequently repeated intention to study the wishes of her people carried the imputation that it was something that her half-sister had failed to do. It was an exercise, it must be said, for which Elizabeth was far better equipped, not just by her history and upbringing, but also by her personal choices and her very public personality. Her connection with her people was solidly based on her ability to make direct contact with them. Unlike her predecessor and indeed her successors, she enjoyed moving among the crowds and milking the adulation. She would go where the crush was thickest and thank them for their love, which served to increase it all the more.

Elizabeth possessed the high skill of appearing to engage with every individual at once. On her arrival in the capital, the whole of London turned out to greet her, and she made it worth their while: "If ever any person had either the gift or the style to win the hearts of her people it was this queen. All her faculties were in motion, and every motion seemed a well-guided action: her eye was set upon one; her ear listened to another; her judgement ran upon a third; to a fourth she addressed her speech".[3] This tour de force was repeated at the procession that preceded her coronation. "Her grace by holding up her hands and merry countenance to such as stood far off, and most tender and gentle language to those that stood nigh . . . did declare herself thankful to receive her peoples' goodwill". When the poor of London shouted out prayers for her peace and prosperity, she answered that she "wished neither prosperity nor safety to herself, which might not be for their common good".[4]

She could claim with some justice that her concern was *their* prosperity and that she cared for it successfully. The contrast with her sister could be extended. In a picture of the family of Henry VIII, painted just over a decade into her reign, Elizabeth is placed to one side of her father, hand in hand with Edward and the images of peace and plenty, while on the other side are Mary and Philip, associated with war and division. This was not just propaganda or empty promises: It was a statement of policy that was already in the process of delivery.

At the end of her reign, she was able to draw on the credit that this had brought her. In the famous Golden Speech to parliament, she had to defuse the anger of her subjects at the spread of monopolies. This she did by renewing the vows of a loving and successful partnership. "And though God hath raised me high, yet this I count the glory of my crown, that I have reigned with your loves. . . . Neither do I desire to live longer days than I may see your prosperity; and that is my only desire. . . . And though you have had, and may have many princes more mighty and wise sitting in this seat, yet you never had nor shall have any that shall be more careful and loving".[5] The familiarity of the speech must not be allowed to diminish its force. The disclaimer of mightiness and wisdom was not entirely disingenuous, for although Elizabeth had actually displayed those qualities in unrivalled fashion, this had depended largely on "reigning with their loves", which was thus indeed the "glory" of her crown. And the fact that she was happy to claim a special capacity to be "careful and loving" shows how specific and significant this stance had been. Recent historians have sought to emphasize the difficulties and discontents that attended her last few years. But this did not alter the uniquely sympathetic relationship that survived throughout. Garrett Mattingly came closest to describing the truth of it: "There has never been anything in history like the forty-five-year-long love affair between Elizabeth Tudor and the people of England".[6] The consequence was momentous. We need not suppose that Elizabeth was deliberately setting out to switch the focus from a dynastic to a national perspective, but that was the effect of her approach, in the context of the time.

In the task of bringing peace and harmony in the religious sphere, she was assisted by her own remarkably undogmatic stance. Although it fell to her in 1559 to institute a formally Protestant religious settlement, it seems that her own views stopped short of outright theological Protestantism. She displayed a notable spirit of compromise. Her preference was for the first settlement of religion made under the aegis of the Duke of Somerset in 1549, with its ambiguity about the meaning of communion. It stressed the nature of the service as a memorial of Christ, but much of the old wording remained, and the more Catholic interpretation that Christ was present in some form was not excluded. In the event, the balance of opinion in the Privy Council and the House of Commons obliged her to accept a more clearly Protestant settlement, but the queen insisted on retaining enough of the latitude of 1549 to offer her loyal Catholic subjects some gesture of inclusion.

Some have thought indeed that she is best regarded as a Henrician Catholic, though Diarmaid McCulloch, probably rightly, rejects this view.[7] Perhaps her position is best described as reflecting the essential character of the indigenous English Reformation, with no pronounced theological bias, but seeking to change the emphasis from an externalist to an evangelist form of worship. Although she did not reject the trappings and observances of ceremonial, she apparently disliked the outright externalist effects of the old religion. Unlike her father, she did not accept the concept of transubstantiation. At her coronation procession, she was content to see a Reformist-inspired tableau heralding her arrival as the end of "superstition" and "ignorance". Her first clear statement on the religious issue came when the bishop officiating at the Christmas service declined to obey her request not to elevate the Host. The queen in turn declined to attend.

By the same token, she showed a commitment to the dissemination of the scriptures in the vernacular, even to the extent of putting herself in danger. Under house arrest at Woodstock under Mary, she asked for an English Bible. This was the essential instrument of Reformed religion, and to demand it was usually regarded by the Catholic authorities as tantamount to heresy. Since she was quite capable of reading the Bible in Latin, Elizabeth's request was presumably a matter of principle. And when, at her coronation procession, an English Bible was handed to her by a child representing Truth, "she kissed it, and afterward applied it to her breast, promising to be a diligent reader thereof".[8]

So Elizabeth's latitudinarian approach was not merely pragmatic. She followed the principles of the core English Reformation to their furthest point, where the rejection of physical effects and the privileging of an intellectual process led towards a tolerationist view. In many ways she epitomized this distinctive trend. She shared not only the theological compromise reflected in Somerset's 1549 settlement but also, quite specifically, the changing conceptual balance that underlay the tolerationist ideas that emerged from the Duke's circle of advisers. Like William Turner and the Discourse, she had come to regard the mind or conscience as a separate and inviolable sphere, which could not be accessed by physical means.[9] In fact, Elizabeth coined the phrase that expresses better than any this crucial psychological change, declaring, famously, that she would make no windows into men's souls. Some of her statements placed her at the cutting edge of tolerationist opinion. She seemed to share the view attributed to leading irenic voices like Richard Hooker and Jacobus Acontius that only God knew the whole truth and that the opinions of men should be treated not as contending truths but just interpretations. As she remarked in 1560 to a Scottish diplomat, "In the sacrament of the altar, some thinks [one] thing, some other, whose judgment is best, God knows".[10]

In practice, this meant that "heresy ceased to be an indictable offence. It was consistently maintained that the queen intended to promote a decent and saintly order in externals, and as consistently that it was not her intention to dictate what a man should believe or to enquire into tender consciences".[11] She made every attempt to honour this principle, even as the political threat of militant Catholicism became manifest. In the wake of the Papal Bull of 1570, which recommended her deposition, she stood steadfastly by her insistence that consciences were not to be invaded. She had a declaration of the queen's proceedings "prepared for general dissemination, and had carefully corrected the text in her own hand, emphasising its liberal doctrines by her amendments". In this statement she drew a direct contrast between the lenient, inclusive nature of her policy and the harsh, rigid partisanship of her predecessor. She promised that as long as her subjects remained outwardly conformable, they would enjoy her accustomed latitude, "without any molestation . . . by way of examination or inquisition of their opinions for their consciences in matters of faith, remitting that to the supreme and singular authority of almighty God, who is the only searcher of hearts".[12] Elizabeth shared and affirmed the distaste that was growing among her subjects for the process of inquisition.

Some contemporaries thought that this kind of restraint could only stem from an absence of beliefs and that Elizabeth actually had no religion at all. A more interesting "secular" implication of Elizabeth's position concerns the way that the queen herself *became* a religion. The cult of Elizabeth was of great importance, but the real force of it has been somewhat obscured. There has been a tendency to treat it as court inspired, in every sense a diversion, a play of chivalry among her entourage, while the wide dissemination of her image is seen as a propaganda exercise to paper over weakness and division. This interpretation is misleading. The truth was rather different. There was no need to induce the circulation of her picture. The demand from "all sorts of subjects both noble and mean" far outstripped supply and led to the problem that inferior copies were being produced to satisfy the market.[13] The cult of Elizabeth arose most crucially from the spontaneous public adoption of the queen as the personification of their inner strength, their prosperity and their prospects.

Already by 1563, the MP Robert Atkinson, expressing his approval of the queen's reluctance to invade men's consciences, went on to hail the general advantages of Elizabeth's approach. "I ween never prince since the Conquest (I speak it without flattery) hath . . . reigned over us in a quieter peace, with more love and less exaction".[14] Atkinson exemplified the success of Elizabeth's policy. He was inclined towards Catholicism and faced harassment from some quarters, but he survived and thrived sufficiently to be able to offer the queen loyal assistance at the time of the Armada. In the intervening decades, the sense grew ever stronger that Elizabeth had inaugurated an age of harmony and plenty, so that by the middle period of her reign it could be said that the kingdom was "never better in worldly peace, in health and body, in abundance of victuals".[15] The statistical evidence for the exceptionally low level of mortality and high level of life expectancy in the middle decades of Elizabeth's reign has already been noted.[16] These benefits were the underlying reason that by the early 1570s the anniversary of her accession was being marked by public celebrations. Again, there was no need for central orchestration. The custom "flowed by a voluntary current all over this realm".[17] Accession Day was soon being celebrated in every parish in England, with bells, bonfires, sermons, open house hospitality, feasting and plays. "Her accession day, 17 November, became the greatest of all national holidays".[18]

In an important sense, it was the *first* truly national holiday. The medieval world had celebrated a multitude of holy days, for many purposes, but none had constituted what was in effect an expression of national self-congratulation. It has been observed, perceptively, that all the devotional energies that had once gone to sustain the veneration of the panoply of Saints were now being directed towards the single person of Elizabeth. These differing forms of adulation had a common feature in the practical motives behind them. And in the provision of health and wealth, the worship of Elizabeth was working beyond all normal expectations. By the golden middle period of her reign, she was delivering greater benefits than the Saints had managed in a millennium.

So the mythology of the cult was secondary to the public approval rating, but it did carry real significance. It had to reflect the circumstances and personal inclinations of the queen, as well as the responses of her people. The celebration of her virginity became important, but this was not in place at the beginning of her rule as has sometimes been supposed.[19] Her reluctance to settle on a husband and thus settle the succession was famously of great concern to her parliaments in the early part of her reign. It seems that although she was prepared to consider marriage, she was equally prepared to rule alone, and it is probably fair to say that "the choice of a husband involved so many contingencies" that she never felt justified in making it.[20] As time went by, the chances of her concluding a fruitful match diminished, but the sense increased that her personal government was not only successful but exceptionally so. A different rationale presented itself.

In the most common allusion, she was celebrated as Astraea, the just virgin, described by Virgil in the Fourth *Eclogue*. Astraea was the last of the immortals, who had abandoned the earth because of the wickedness of mankind and inhabited the heavens as the constellation Virgo. Her return to earth was foretold and would inaugurate a golden age, bringing peace, plenty and eternal springtime.[21] The analogy seemed fairly precise. The prosperity of Elizabeth's realm was undoubted. Her even-handed justice was manifest to all. She was indeed distinctive as the Virgin Queen, which might be considered as a necessary freedom to enable her to exercise her beneficial power. She loved only her people. She was not encumbered by a husband or by the traditional princely pursuit of personal glory. Her commitment was to the aspirations of her subjects. As luck would have it, Astraea was also synonymous with the moon, one of the properties of which was to govern the tides. This became a popular metaphor with men like Sir Walter Raleigh, who wished to foster the idea of Elizabeth as mistress of an empire of the sea.[22] Thus the queen offered a clean sheet, on which the nation could sketch in the vision that had opened up before it. Seeing her in this role became a standard perception quite early in the reign, and it was depicted with remarkable power and precision. In a portrait by Massys in 1579, she carried a sieve in reference to the Vestal Virgin Tuccia, whose purity enabled her to carry water in that unlikely receptacle. On a pillar behind Elizabeth was the story of Aeneas, who had forsaken Dido and founded Rome, as Elizabeth must avoid entanglements, leaving her sailors and traders free to roam the seas in search of empire. In the background was a globe, on which, from a sunlit England, ships were setting out towards the west.

The significance of these images of Elizabeth has perhaps not been fully understood. We see it better if we contrast it directly with the standard depiction of the male prince. Typically he appears on a prancing horse, in armour, with sword at the ready. This is the personification of war for war's sake. This is what king's do. He is off somewhere in pursuit of personal glory or the assertion of dynastic rights. It is rather different for Elizabeth. She would probably not sit too well on a prancing horse, and armour might not sit too well on her, but she has a kind of natural poise when portrayed standing atop a globe of the world. And although it may give little comfort to the feminist tendency among her biographers, this is precisely because she is static. In a portrait by Marcus Gheeraerts the Younger, she stands on and secures the island as a sound and solid base, from which her people may advance across the oceans, which stretch away beyond.

The vision peaked in the Armada portrait. The great fleet had arrived in all its pomp on one side of the queen and had been dispatched in ruin and disarray on the other. Again Elizabeth stood for the physical integrity of the island. The Virgin Queen reflected the inviolability of her realm. She remained quite static. She was the rock on which the Armada had broken. Unlike her male counterparts, she had no need to go anywhere in the pursuit of her glory. But she personified the island, and she rested her hand on the globe. The Spanish grip had been loosened. The surrounding seas had been cleared. The oceans of the world were presented to her people, where once the triumphant feudal lord would have granted territories to his lieutenants.

It had taken a while for the great climactic confrontation with Spain to materialize, but the elements of conflict were always there as a simple reflection of circumstances. At first, Elizabeth's ambiguous position on questions of the faith led some Catholic authorities to hope that she might be persuaded back into the fold, but it gradually became clear that England was emerging, politically if not theologically, as the champion of European Protestantism, and by 1570 the Papacy had concluded that it need not defer ordering Elizabeth's deposition.

The basic clash of interests was always evident in the battle for trade and empire. In 1561, Secretary of State Cecil issued a declaration of intent by repudiating the Spanish claim to a monopoly of trade in the New World, originally granted to them by the Pope. This was the field in which the provocation of Spain had to be risked, and to some extent sought, from the beginning. In 1562, John Hawkins set forth to carve out a share in the lucrative trade of African slaves to the West Indies. At first the Spanish planters accepted this source of supply, and high profits accrued for all. This tempted the queen to take an interest in the second voyage. But by the third, the Spanish authorities had decided to defend their trading areas more vigorously, and the venture concluded with a pitched battle. "It was impossible any longer to disguise the fact that although England and Spain were nominally at peace in Europe, a state of war prevailed 'beyond the line' ".[23] The irreducible foundation of the contest was that the English people had set a course to supplant the trading and imperial power of Spain.

The arrival in 1567 of a powerful Spanish army to suppress the Dutch Revolt showed that the contest for empire implied a new geopolitical position for the English state in its relationship with the Continent. The presence of a Spanish force across the channel posed a threat to the English coast that could not be ignored. Thus appeared the outline of the shape that British foreign policy was to assume for the next 350 years. The core proposition was to obstruct the emergence of an overweening military power across the channel, while developing and extending the ability of Britain to control the seas. Historians who suppose that the agenda might have been different are underestimating the extent to which it dictated itself, given the force of the imperial project.

The new strategic balance was further delineated following an incident in 1568, when ships carrying money to pay Spanish troops were attacked by Huguenot pirates and forced to put into Southampton and Plymouth. The English hosts were unable to restrain their natural instincts, and the treasure was commandeered. The Spanish were understandably incensed, and all English property in Spain and the Netherlands was confiscated. Elizabeth retaliated, and a trade war ensued, severing commercial relations for five years. The traditional trade route through Antwerp never really recovered. It became clear that in view of the underlying enmity with Spain, the London–Antwerp axis needed to be supplemented or replaced. Ironically, the Spanish action had provided the final, crucial incentive for English naval forces to range with complete freedom across the oceans. "With the collapse of the Antwerp market in 1569–73, the English cloth trade was freed from its dependence on the House of Burgundy. The merchants had to seek out new marts all round the coast of the continent".[24] This they did, with Royal Navy escorts.

As the contest at sea developed, some small and, on the queen's part, hesitant steps were taken to aid the rebellious Dutch Provinces and impede Spain's attempts to reconquer them. Elizabeth countenanced volunteer expeditions and sent a little money, though often by clandestine means. More direct action could be deferred because during the 1570s the Spanish crown was often in financial straits, coping with English privateers in the Caribbean and the Turks in the Mediterranean. The unpaid Spanish troops in the Netherlands staged a series of mutinies, and the newly independent United Provinces survived.

By the early 1580s, however, the situation had changed significantly. Spain had completed the conquest of Portugal, and the conflict with the Turks had ended.[25] The flow of money to the army in the Netherlands became steady, and the Duke of Parma embarked on what threatened to be a systematic reconquest. In England, Catholic plotting intensified. The recovery of both Scotland and England by a Habsburg/Guise alliance was seriously projected. In 1584, the danger took definite form in the Throckmorton Plot to murder the queen. The Spanish authorities were closely implicated in this, and the ambassador was expelled. With the Dutch position further weakened by the death of William of Orange, action could no longer be avoided. In October 1584, the Privy Council determined on a war policy—there would be an attack on the Spanish West Indies, and an expeditionary force would be sent to help defend the independent Dutch provinces.

This was a decisive moment in establishing a new national perspective in the field of foreign affairs. Wallace MacCaffrey noted in an important essay that when Spain was named as the enemy in 1585, royal policy was catching up with what had been the clear and dominant sentiment of the political nation since the middle of the century. But this was not simply the crown settling on a policy that happened to be in line with the preferences of its people. The process entailed a qualitative change in the way that foreign policy was conceived and articulated. It no longer revolved around the personal appeal to feudal loyalty or the imperious summons to aid the king in asserting his rights. The war aims were of a different nature and were presented in a different form. They were carefully laid before parliament, recognizing the fact that they reflected not the dynastic designs of the crown but a set of independently defined national priorities. "In these pre-war parliaments one sees the fulfilment of the shift which had been taking place for decades past, away from a policy built on a dynastic view of foreign relations to one orientated on national interests".[26] The crucial factor was that the strongest voice in defining the agenda was that of the political nation itself.

No doubt they were gratified that their view of the Spanish as invasive and tyrannical seemed to be borne out by the appearance of the Invincible Armada over the horizon. The question now was whether the national project of supplanting Spanish sea power could be sustained in the face of the collected force of it. The answer was yes, if only just. The fact that the defeat of the Armada was less than clear-cut, its destruction owing more to the weather than to English naval action, has given recent historians a pretext to downplay the story of Elizabethan success. Even one of those most open to the idea of a popular foreign policy, Tom Cogswell, sees James I as being in the shadow of the "myth" of his predecessor's triumphs and thinks that in actuality her war against Spain "had not proved an outstanding success". He then convincingly contradicts himself by adding that, "admittedly she secured the independence of the new Dutch Republic and the French monarchy".[27] When we recall that she also secured the independence of England itself against the greatest military threat since the Norman Conquest and stabilized the future of the Protestant religion in Europe for some decades, we may wonder what on earth Elizabeth would have had to do to qualify as a success in the eyes of modern historians.

The recent determination to undermine the reputation of the Elizabethan age is one of the most damaging distortions of revisionist history. If it stemmed from a reluctance to glorify the tradition of military success, it would be more forgivable. But there is a far less rational process behind it. The revisionist generation has been largely driven by an excessive fear of "teleology", or interpreting history in terms of a given end. The phobia stemmed mainly from a dislike of Marxist theory, which sought to predict or influence the future in a favoured direction. To increase the power of this, Marx had treated it as a scientific projection, and his detractors talked as if he had somehow turned history into an "inevitable process". They applied the same critique to the Whig thesis of constitutional advance and concluded that any concept of progression should be disallowed.[28] But to approve or disapprove of past events is not necessarily to regard them as inevitable or to misinterpret and invent them in the light of the present. It may be, on the contrary, a clear recognition of the most powerful evidence of change, as well as the most obvious and only way in which the past can move, via the "trace of what precedes and follows".[29] To talk of anything that presages the future as a preconception is to conceal the force of coherent change. It is wrong to suppose that when one generation takes a positive, formative view of events in another, this is necessarily mistaken or "mythical". On the contrary, it can reflect the real dynamic of history.

The discomforting of Spain was not imagined by the political nation but recollected as an event of real significance on which they wished to build. It created not a myth but a vision. Or, to be more precise, it *substantiated* a vision. There was a positive importance in the fact that it had been achieved without conclusive military action. The island had shown that it could defend *itself*, that it was indeed a "fortress built by nature for herself". The most interesting aspects of the campaign were its non-military features. It reflected a new relationship with the sea and the elements. It was the first major sea battle to be fought not by galleys propelled by oars but sailing ships dependent on the wind and their ability to use it. The English ships had proved more mobile, and the English sailors more in command of the new skills that were required. As Colin Martin has said, English warships of the latest design "were especially strong and stable . . . with a sailing performance far superior to that of the best of their rivals".[30] If the weather appeared to be on their side, it was in part because they were more at ease with it. The sailors celebrated their ships as being more "weatherly". They could sail nearer the wind than their opponents. The crucial indicator for the prospects of English naval activity was not their military success but the fact that they felt peculiarly at home on the sea.

It also needs to be recognized that the underlying issue was more about mercantile enterprise than military ambition. This was reflected in the make-up of the English fleet. The Royal Navy had been created in the reign of Henry VIII and was well maintained in Elizabeth's, but it remained small at about forty ships. There was, however, an extra dimension, in that as many as seven hundred merchant vessels were available to be requisitioned when the need arose, and they were apparently of suitable and advanced design. Thus the Newfoundland fishing enterprise that operated from the West Country ports was called off for a season while the fishermen and their ships served to complete Drake's Western Squadron. They were reciprocating a naval action of five years before, which had swept the Newfoundland waters clear of Spanish ships and allowed the fishing venture to thrive. This was a microcosm of the national project.

A sequence of events that occurred a year after the Armada showed rather more clearly the scope and nature of the ambition, as well as the reasons why it could now be pursued with complete self-confidence. Drake proposed, with the help of Norris and his men, to attack Lisbon, with the aim of inspiring a regime change in Portugal. This, it was hoped, would not only embarrass the Spanish crown but also have the considerable benefit of opening up the Portuguese Empire to English merchants. Thus was displayed the full scale of the prospect that the political nation saw unfolding before it. Drake's plan was a little overambitious, but it did achieve some success. He disrupted the progress of the treasure fleet and helped to provoke another mutiny among Parma's unpaid soldiers. The English ships then intercepted a French and Hanseatic fleet, which was on its way to re-equip the Spanish navy. They took it into custody and redirected it to Plymouth, where they no doubt felt the supplies would be put to better use.

This was the real lesson of the Armada campaign. As John Guy concluded, "Elizabeth's navy had extended her control over the coastline of the British Isles and the waters of north-west Europe".[31] This completed the concept of the defensible island and the serviceable surrounding sea. It was the platform on which the English people could perceive an enticing and prosperous future stretching out before them. The sense of a new capacity to move with complete freedom across the oceans was crucial. The sea covered the greater part of the earth's surface. To establish command of that would bring power and wealth beyond measure. New technologies of boat building and navigation had created the means by which it could be done. And it was of great significance that this was the project not so much of the crown but rather of the political nation itself.

The assurance of the physical integrity and inviolability of the island was seminal to the developing national awareness. So too was the recognition of especially favourable qualities or circumstances, which gave the national land a distinctive character to be defended. A few years after the Armada campaign, Shakespeare wrote Richard II and included a famous speech celebrating not just the glory of the Elizabethan nation but its happy establishment and proven viability. The passage is frankly rather excessive, as a description of the kingdom that the speaker, John of Gaunt, feared that the effete Richard II was about to discredit. But it carries enormous power as a celebration of the contemporary Tudor nation. The real historical significance of John of Gaunt was as the main prop of the Tudors' somewhat insubstantial claim to the throne. So here was aged Gaunt bestowing a posthumous blessing on the happy and consolidated land that had emerged under the rule of his successors. And here was Shakespeare giving resounding voice to the general sense of satisfaction.

> This other Eden, demi-paradise,
> This fortress built by nature for herself
> Against infection and the hand of war,
> This happy breed of men, this little world,
> This precious stone set in the silver sea,
> Which serves it in the office of a wall,
> Or as a moat defensive to a house,
> Against the envy of less happier lands.[32]

The play "took London by storm when it first appeared. We have it on the authority of Queen Elizabeth herself that it was acted forty times, an unusually long run for an Elizabethan play".[33] Those who listened to this speech for the first time thought not of Plantagenet England but of Tudor England, and they thought of England in a way that would not have been possible in previous ages. It had become a self-possessed entity: "this dear, dear land . . . bound in with the triumphant sea". The sea defined and secured it. The idea of a defence "against infection" is of great interest. It is unlikely that this was the assumption of a barrier against incoming disease, though it did perhaps involve an awareness of the unusually good health that Elizabeth's realm was enjoying at that moment, in the golden middle years of plentiful harvests and high life expectation. But most probably it indicated a desire that the special characteristics that went to make up the new nation should not be compromised. Its identity was sacrosanct.

Thus it appeared vital to sustain "this little world" as something complete and separate. Happily, nature had also provided the means for this. In fact, in Shakespeare's imagery we can envisage the form in which the physical land had turned into a national land. The island had taken on a metaphor—it had become a fortress. All the actual castles that had sustained the sectional and regional powers within the kingdom were gone. The island itself was now the castle, just as the land was now a conceptual unity. And the surrounding seas were no longer the negation of the land or the unwanted waters running off into the drainage furrows. The sea served the new national land in a much more positive sense, for all its needs. It was the moat and wall, the barrier to invasion and the guarantee of the nation's integrity. It was also—and here was the beauty of the thing—a free way out for the nation's sailors to plough their routes and ply their trades, across the oceans on the way to future prosperity.

In these same years, following the defeat of the Armada, Richard Hakluyt produced the seminal text of the national project, *The Principal Navigations, Voyages, Traffics and Discoveries of the English Nation*. In the very title we can hear the word "nation" acquiring its new meaning, as the physical land of the English race took on a distinct political and economic character. An emergent national land was gathering itself within its own independent and coordinated shape, galvanized by the vision of a trading empire to be won by exploration and enterprise beyond its shores.

The scale of Hakluyt's work was as grand as the project itself. It was a massive book, describing ventures to all parts of the world. The global reach was the essence of the concept. The dedication of the first edition declared that in Elizabeth's reign the English people had shown themselves peculiarly well suited to this mission and that "in searching the most opposite corners and quarters of the world, and . . . in compassing the vast globe more than once, have excelled all the nations and peoples of the earth".[34] It has been said that Hakluyt established the vocabulary of empire. He gave a definitive voice to the vision. But we should note again that it was a mercantile vision.

"Which of the kings of this land before Her Majesty had their banners ever seen in the Caspian Sea? Which of them hath ever dealt with the Emperor of Persia . . . and obtained for her merchants large and loving privileges? Who ever found English consuls and agents at Tripolis in Syria, at Aleppo, at Babylon, at Balsara, and which is more, who ever heard of Englishmen at Goa before now?"[35] He envisaged an empire of commerce. "His interest lay in the extension of markets, in the development of trade, in the utilization of products of new territories for the benefit of home markets".[36] Or as he put it in the dedication of the second edition to Cecil, "our chief desire is to find out ample vent of our woollen cloth". It was the ultimate expression of the trend noted above as a crucial factor in determining the definition of a national land—that is, the emergence from the mid-sixteenth century onward of commentaries that identified the island as a commercial unit and focused on the question of how the processes of trade could be made to serve it. Just as the extended range and coordination of internal marketing bound the nation together, so the broader scope and awareness of external commerce defined its outward relations.

The whole world was now a potential market. The Elizabethan nation had come to believe that there was nowhere that the sea could carry them that could not be "opened" to their profit and advantage. Hakluyt sought to promote the exploration and exploitation of the Americas as especially profitable. But if that was the area where the greatest gains were to be expected, it was also the place where the power of Spain comprised the most considerable obstacle.

NOTES

1. J.E. Neale, *Elizabeth I and Her Parliaments*, (London 1953), I, pp. 42–3, 50.
2. Ibid., pp, 42–3.
3. John Strype, *Annals of the Reformation*, (London 1824), I, p. 15; John Hayward, *Annals of the First Four Years of Queen Elizabeth*, ed. J. Bruce, Camden Society VII, (1840), pp. 6–7.
4. James Osborn, *The Queen's Majesty's passage through the City of London to Westminster the day before her Coronation*, ed. J.E. Neale, (New Haven 1960), pp. 27–9, 44.
5. Ibid., II, pp. 389–91.
6. G. Mattingly, *The Defeat of the Spanish Armada*, (London 1983), p. 22.
7. D. MacCulloch, *Tudor Church Militant*, (London 1999), p. 186.
8. James Osborn, *The Queen's Majesty's passage through the City of London to Westminster the day before her Coronation*, pp. 38, 54.
9. Ibid., pp. 38–40.
10. A. Somerset, *Elizabeth I*, (London 1991), p. 80.
11. M.J. Tooley, "Political Thought and the Theory and Practice of Toleration", *Cambridge Modern History*, III, pp. 491–2.
12. J.E. Neale, *Elizabeth I and Her Parliaments*, I, p. 191–2.
13. A. Somerset, *Elizabeth I*, p. 355.
14. Ibid., I, p. 118.

15. The Bishop of Salisbury, cited in S. Schama, *History of Britain*, (London 2000), p. 369.
16. See above, p. 90.
17. Thomas Holland, cited in R. Strong, *The Cult of Elizabeth*, (Pimlico 1999), p. 118.
18. S. Schama, *History of Britain*, p. 372.
19. S. Doran, *Monarchy and Matrimony: The Courtship of Elizabeth I*, (London 1996).
20. Ilona Bell, *Elizabeth I: The Voice of a Monarch*, (Basingstoke 2010), p. 5.
21. R. Strong, *The Cult of Elizabeth*, pp. 16, 114.
22. A. Somerset, *Elizabeth I*, pp. 354–5.
23. J.B. Black, *The Reign of Elizabeth*, (Oxford 1965), pp. 123–5.
24. G.D. Ramsey, "The Foreign Policy of Elizabeth I", in *The Reign of Elizabeth I*, ed. C. Haigh, (London 1984), p. 167.
25. F. Braudel, *The Mediterranean*, (London 1973), II, pp. 1184–5.
26. W. MacCaffrey, "Parliament and Foreign Policy", in *The Parliaments of Elizabethan England*, ed. D. Dean and N. Jones, (Oxford 1990), p. 80.
27. T. Cogswell, *The Blessed Revolution*, (Cambridge 1989), pp. 12–4.
28. For instance, B. Coward, *The Stuart Age*, (London 1980), pp. 160–1.
29. Ibid., pp. 7–8.
30. C. Martin and G. Parker, *The Spanish Armada*, (Harmondsworth 1988), p. 54.
31. J. Guy, *Tudor England*, (Oxford 1988), p. 342.
32. Shakespeare, *Richard II*, act II, scene I.
33. *Richard II*, ed. John Dover Wilson, (Cambridge 1968), Introduction, p. ix.
34. Dedication to Sir Thomas Walsingham, R. Hakluyt, *The Principal Navigations, Voyages, Traffics and Discoveries of the English Nation*, (London 1589).
35. R. Hakluyt, *The Principal Navigations, Voyages, Traffics and Discoveries of the English Nation*, (London 1599). Everyman edition, (London 1907), I, p. 3.
36. J.D. Black, *The Reign of Elizabeth*, (Oxford 1965), p. 319.

14 The Foreign Foreign Policy of James I

The emergence of a self-defining popular concept of foreign policy was the most obvious reflection of the physical land turning into a national land, a self-conscious political unit with a set of clear and distinctive priorities to be asserted and defended against others. This involved a change in the way that the public interest was defined. In times past, the unitary interest of the kingdom had been determined by the concerns and position of the king. If it had not been for the clashing dynastic conflicts of their princes and nobles, the interests of the peoples of medieval Europe would have been essentially non-competitive and indistinguishable. Now the people of England were beginning to identify their own national imperatives.

So the assertion of a freestanding public interest in foreign affairs posed a challenge to the prerogatives of the crown and created the most conspicuous cause of dissension in the controversies between the political nation and the king in early Stuart England. It became a conflict of interests over matters of great substance, including the rights of public finance, and it proved to be beyond conciliation. This is explained in part by the force of the popular concept of foreign policy. In contradicting it, James I was not in the shadow of a myth; he was standing in the way of an established national project, which had lately proved its viability and seemed to indicate a clear future path. James was placing himself in opposition to the kingdom's own developed view of its priorities.

In the prejudice against the force of change and long-term causes, there has been a tendency in modern historiography to portray the reign of James as a time of balance and normality. This overlooks the critical differences of political practice and principle that were coming to divide king and people. I have described, here and elsewhere, the emergence of two opposing definitions of the rights and requirements of government—the developing contest between the prerogative and representative forms of administration, between "patent" and "parliament".[1] In some respects this was beyond resolution. It was not really possible for James to moderate parliament's fixed determination to put an end to discretionary customs dues, not unless he was prepared to surrender important elements of his prerogative. In the field of foreign affairs, however, he had more leeway. It would have been conceivable and indeed logical for him to adopt a military policy in line with the dominant sentiments of his subjects. Why did he choose to do the opposite?

It is not difficult to identify elements of James's background and beliefs that gave him different priorities from those of his new subjects. In Scotland, he had been obliged to govern in the face of violent noble factionalism and a strong-willed Kirk that had been established by popular revolution, was structured independently of the crown, and entertained theories that justified resistance to an ungodly monarch. He was left with a horror of anything that savoured of rebellion or "popularity", and he wrote tracts specifically to warn against it. The defence of royal authority became his guiding light. So he was more inclined to reject the force of popular opinion than to accommodate it. He could not accept the views of his people as reflecting the real needs and strengths of the kingdom, as Elizabeth appears to have done. This is the essential clue to James's approach—for various reasons, he was not able to develop a concept of an objectively defined national interest. He never got beyond the traditional view that the personal interests of the king decided the unitary needs of his realm.

James was frequently exercised to deflect the developing national perspectives of parliament. In 1609 he lectured the Commons on the acceptable limits of their role: "Ye should only meddle with such grievances as yourselves do know had need of reformation, or were informed thereof in the countries you serve . . . do not meddle with the main points of government, that is my craft: to meddle with it were to lessen me".[2] So the particular grievances that MPs voiced on behalf of their localities were not expected to include a view of foreign policy any more than they had been expected to include a denial of prerogative impositions. In assuming sole initiative in foreign affairs, James was claiming no more than his traditional right. If confirmation was required, it could be found in the words of Sir Thomas Smith. "The prince has absolutely in his power the authority of war and peace, to defy what prince it shall please him . . . and again to reconcile himself or enter into league or trust with him at his pleasure or the advice only of his privy council".[3] The problem was that there now existed in the kingdom an independent concept of foreign policy, and it was just as absolute a perception as the king's view of his personal power.

The parliamentary stance was certainly subversive. In attempting, as they would, to force a change and indeed a reversal of the king's foreign policy, they would be infringing the royal prerogative just as deeply as in their more direct attempts to eliminate prerogative rights, like the power of impositions. And parliament's assertiveness only reinforced James's aversion to "popularity". His choices and responses in foreign affairs were always conditioned largely by his desire to uphold monarchical rights and his abhorrence of anything that seemed to involve or encourage their usurpation, in England or elsewhere. In effect, his approach became as much a way of dealing with political challenges at home as with military challenges abroad.

James saw a positive political purpose in seeking to remain free of parliamentary influence.

The touchstone was his attitude to Spain. It was the settled, leading view among his subjects that Spain was the natural enemy, the antithesis against which the national interest was in every sense to be measured and asserted. But James was apparently not able to formulate or embrace a "national" definition of foreign policy, and in a sense he did not possess a basis for so doing. Although many of his subjects retained a defining recollection of the struggle for national survival against Spain, James had no direct reason to share it. What he did retain was an anxiety to sustain the rights of kings, so it was more natural for him to embrace Spain as the strong monarchy most to be admired and as the ally most avidly to be courted. The Spanish methods of government that the English political nation classified as "tyrannical" were to James more the mark of a properly sovereign "free" monarchy.

Being determined in most respects by an anxiety to assert monarchical rights, James's policy was, in his own terms, quite well coordinated. An alliance with Spain would address his domestic problems, as well as his foreign concerns. By that course he would be associating himself with a model hereditary monarchy, which did not have to cope with insubordinate representative assemblies and was so rich that it might help him to bypass the one that he had found in England. A marriage treaty with Spain would bring him more money than he could ever hope to get from parliamentary subsidies, and this in turn would free him of the need to suffer the pretensions of the House of Commons. So his policy was followed not only without reference to the predominant desires of his people, but also in some significant degree so that he could avoid making application for their support or having to listen to their views.

For the first decade and a half of his reign, the sharpness of this divide was concealed by the fact that the kingdoms of Europe were militarily and financially exhausted, and James could exercise his supposed talents as a peacemaker without any great problem. He fondly imagined that he had established a settled repose in the international field. To complete this orderly vision, James's favoured initiative was to seek a Spanish bride for his heir, first for Henry, then for Charles. When he published his intention for a Spanish match for Charles, the commissioners detailed to consider the treaty proposals gave a reply, which "veiled a repugnance to the proposed marriage".[4] The king proceeded regardless. He would be reprising the dynastic connection that had been so vigorously denounced by public opinion in the reign of Mary Tudor, and great events since then had only served to make the policy seem even less appropriate now. But if the political nation saw opposition to Spain as an imperative, there was nothing in James's background or beliefs that gave him reason to agree. He had no concept of an objectively defined national policy, and every personal consideration inclined him towards Spain. An alliance with the great Catholic power would extend the climate of peace that James believed he was nurturing in Europe, and dynastically Spain was a very good catch. Moreover, the expected dowry would assist him in ignoring the unruly demands of his subjects. The Spanish match "allowed James to contemplate life without parliamentary subsidies".[5]

The confrontation over the crown of Bohemia in 1618, which proved to be the beginning of the Thirty Years War, did not deflect James from his chosen allegiance, even though it brought the interests of his son-in-law Frederick of the Palatinate and the Protestants of Bohemia into direct conflict with the Habsburg powers. As always, James's responses were governed primarily by his horror of rebellion. "From a war in support of revolt or usurpation he shrank as from the plague".[6] Consequently, he "celebrated his son-in-law's success by officially denying to every sovereign in Europe that he had countenanced or even known of the project".[7] Glyn Redworth confirms that James "never approved of what he called Frederick's usurpation of the de facto Habsburg rights in Bohemia".[8] The Spaniards persuaded James to attempt a peace initiative. This played to James's self-image, but it was never a serious proposition, and the suggestion was really just a ploy to "divert him from taking any action".[9]

Habsburg power was duly reasserted in Bohemia, and it was little secret that Spain and the Empire would seize the opportunity to attack the Palatinate, which was of strategic importance as a staging post between Northern Italy and the Netherlands. Since the Palatinate was Frederick's patrimony, the Habsburgs would now be the usurpers. But this did not sway James from his devotion to Spain as the model of legitimate monarchy. He allowed himself to be persuaded by the Spanish argument that the root of the aggression had been Frederick's attempt on the crown of Bohemia and that the threat to the Palatinate would apply the necessary leverage to restore good order. James readily subscribed to this line. "His reaction to the invasion of the Rhineland was to put pressure on Frederick to renounce the Bohemian throne".[10] James's obeisance to Spanish power suggested to him that even though they had participated in annexing the Palatinate, an alliance with them was still the best chance of getting it back. The crisis thus "served only to wed James more firmly to the idea of a Spanish match".[11] On the eve of the invasion, he was at pains to assure Gondomar of his fidelity to Spain. "I give you my word as a king, as a gentleman, as a Christian, and as an honest man, that I wish to marry my son to no one but your master's daughter, and that I desire no alliance but with Spain".[12]

That much was clear. The complication was that the predominant view in the English political nation ran on precisely the opposite lines—they wished to fight no one but Spain. Even before parliament met, the strength of public opinion was making itself apparent. Taverns, streets and markets echoed with debate of the crisis and with condemnations of Spanish and Catholic militancy. Tom Cogswell has described how the latest international news was disseminated through a network of information centred on St Paul's in London.[13] Gondomar reported to Madrid that he could have sent an armload of seditious pamphlets every week if he had sufficient couriers. The most striking of the pamphlets was by the Norwich divine Thomas Scot, carrying the self-explanatory title "Vox Populi". This offered an extraordinarily graphic description of a king and people completely at odds over the appropriate course of foreign policy. It depicted an English court that "gaped wide for Spanish gold", with a king that "extremely hunts after peace . . . as that for it he will suffer anything". And it suggested that although it was possible that the kingdom might be rallied by a parliament, "the king will never endure parliament again but will suffer absolute want, rather than receive conditional aid from his subjects".[14]

The perception that parliamentary aid would be "conditional" clearly took on an added importance when the political nation had developed such a powerful and prescriptive view of what the policy should be. James's instinctive assessment that this was to be resisted was quite judicious from the perspective of maintaining a "free" monarchy. But his arrangements for availing himself of the Spanish model were not yet sufficiently advanced to enable him to bypass the possibility of parliament's financial support. So the stage was set for the first direct confrontation between a House of Commons, who believed that national policy should take the form indicated by their collective representative judgement, and a king, who was determined to go his own way and maintain his rights.

James had attempted to suppress public debate with a proclamation in 1620 forbidding "licentious passage of lavish discourse and bold censure in matters of state". When this failed to have the desired effect, he issued another, noting that "the inordinate liberty of irreverent speech doth daily more and more increase".[15] A main burden of the agitation was that there ought to be a parliament. I have described elsewhere a typical episode in which the citizens of Bridgewater engaged in heated discussions of the foreign situation, with the central supposition that a parliament was required to take order for it.[16] Robert Zaller went so far as to say that "the country was determined to force a parliament".[17] There was certainly a general desire for an assembly. The king would have preferred to resist the pressure. But parliament's representative and fiscal strength meant that if there were to be any response at all to the Palatinate invasion, the Houses would have to be called.

The king indicated that he would go to war for the Palatinate if peace talks proved unsuccessful, but in James's mouth the ultimatum did not carry conviction. In his opening speech to parliament, he said that he needed to be prepared for war in order to strengthen his hand in the peace negotiations. He went on to betray the fact that the Spanish match remained his only real policy. His assurances that the marriage would not involve concessions to Catholicism showed that he was still putting his faith in the entente with Spain.[18] The Commons saw little enough evidence of the forthright anti-Spanish policy that they favoured. Pending that, they decided to vote the king just two subsidies to cover his immediate contingencies, rather than as a provision for a war. James was said to be delighted with the gift, having in effect received unconditional supply, without committing himself to action. But it boded less hopefully for royal finances in the future. The fact that the grant was specifically *not* intended to cover a war was the first sign of what would become the settled stance of the House: They would give substantial supply only if they were getting exactly the foreign policy that they wanted.

When the king made overtures for a further subsidy, with no war policy in place, the Commons failed to respond. Gardiner thought that this was the probable reason that the king dismissed the assembly at the beginning of June, with their legislative programme uncompleted.[19] It certainly became the settled impression that early Stuart governments had no positive interest in parliament except for the possibility of supply. MPs, however, were distraught at the aborted session. Richard Wynn wrote to his father of their unexpected dismissal "without the effecting of any of our businesses". He also underlined what they had wanted the king to do in the military field. "All the princes of Christendom are arming, yet we are not moved with all this. I pray God keep us, for we were never more in need of his help".[20] Before departure, however, the Commons did manage to produce a Declaration making their position clear and offering the king some inducement to recall them. They said that if perchance the peace negotiations for the return of the Palatinate should fail and the king should decide on a more forceful approach, they would be ready to provide supply. The Declaration was well received in the constituencies, by such as the diarist Walter Yonge, as the best hope of instigating a forward policy.[21]

During the summer, it became painfully obvious that James's peace negotiations had been unavailing, as the Habsburg armies completed the conquest of the Palatinate. It was clear that some semblance of a more forceful response was indicated, and parliament was recalled. The Commons would have to be allowed to debate the issue if they were to advance supply. But James had not abandoned the idea of the Spanish entente. It remained his favoured approach, and he warned MPs against putting forward their own no doubt very different view. They were told to restrict their comments to the financial side of the question and to avoid any analysis of policy. James stressed that there must be no invectives against particular targets, since this kind of freedom of opinion was not acceptable to him.[22]

Given the importance of the issue and the strength of their views, the Commons were unlikely to observe these traditional constraints. They believed that they knew the line of policy in which the vital interests of the nation lay, and they perceived that it was not being given due prominence. The king had offered no indication of any new approach of his own and preferred to be absent at Newmarket on the occasion of the Commons' debate. As so often, James's actions, or lack of them, served to bring about the very circumstance that he was seeking to avoid. On 26 November 1621, the House of Commons settled down for the first time to formulate a foreign policy that they could regard as being in line with the real strengths and requirements of the nation that they represented.

What they had to say was powerful and consistent and possessed in their own minds an irresistible logic. Almost all the recorded speakers thought that the war effort should be directed against Spain. Sir Edward Giles exemplified the sentiment, declaring that their preparations should not be focused just on the Palatinate as such but rather on resuming the essential conflict with Spain. He issued the authentic and proven battle cry, to "take Elizabeth's course . . . find his Indies".[23] Thomas Crew, a leading voice in these affairs, echoed Giles's emphasis on opening the Indies and summed up the mood of the House. "We all see which way we incline".[24] The stress on "opening" the West Indies was not just a general identification of Spanish interests but a specific, selected line of attack. This was what the Commons' understanding of naval and trading activities had identified as offering the best chance of profit and success. It was what John Pym called "the most facile and hopeful way for this kingdom to prevail against the Spaniard".[25] Thomas Crew outlined how such a war could be fought and maintained, and the debate "reached its crescendo in a violent speech by Coke, who throwing off all restraint, poured forth vituperation against Spain".[26] MPs of lesser status who were not recorded in debate nevertheless expressed the same clear and powerful views. John Prowse and Ignatious Jourdain of Exeter wrote letters to their constituencies, condemning Spain as the "invertirable enemy" and taking hope from the arrival of some Dutch envoys, though they could not give assurance as to how the king had received them.[27]

Thomas Crew concluded with a thinly disguised demand that Spain should be named as the enemy, and Prince Charles should be married to someone of his own religion. "How glad it would make us, and willing to give".[28] The following day they proposed to seek explicitly to have Spain named as the enemy, and when the Treasurer of the Duchy of Lancaster tried to obstruct the move, "he was quickly crossed" by William Nyell of Dartmouth "resorting to the former question against whom we should fight". Nyell wanted a seaborne assault on the worldwide interests of Spain and gave assurance that his constituents would provide support for such a war.[29] These were clear expressions of what had become the settled stance of the Commons. They would only give maximum supply if they were getting precisely the policy that they desired. In this same spirit, Sir Nathaniel Rich pointed out that their commitment to extra supply at the recess had been made on the condition that James would undertake the kind of war that they envisaged.[30] So they voted just one subsidy now, and they drafted a petition asking for Spain to be named as the enemy and for the prince to be married to a Protestant.

The force and unanimity of the anti-Spanish perspective, both in the Commons and across the constituencies, is very striking. In this, we see the process by which a national consensus in foreign policy was constructed. A set of public imperatives was established by common experience and confirmed through the uniformly reflective representative function of parliament. The sense of a national land with distinctive interests was consolidated, as the island community identified, by means of collective, independent assessment, its most advantageous relationship to the outside world.

Historians seeking to minimize the force of popular opinion have strained to create a restrictive view. Conrad Russell supposed that "only [Sir Nathaniel] Rich, with a proposal that any league the king might enter into should be confirmed by parliament, had offered any move in the direction of parliamentary control of foreign policy".[31] In truth, Rich's idea, though certainly radical, was a mere technicality in the context of the general stance of the Commons. The truly revolutionary aspect was that the Commons, as a body, were not happy to endorse the one league into which the king wished to enter—they were seeking to push James in their own chosen direction. This was a groundbreaking attempt to determine the shape of the kingdom's foreign policy, in contradistinction to that of the king.

Their position proved, of course, to be completely unacceptable to James. As he rightly perceived, they were attempting to subvert his personal intentions and invading a prerogative sphere in which they had no right to meddle. He told them that he had only invited them to debate the issue in terms of how they would supply him, not as a pretext to advance an alternative view of world affairs. This, he said, was to usurp his place and assume the principal power in the state.[32] This was undoubtedly the case. But the Commons were not inclined to withdraw. They saw the force of these imperatives too clearly for that. Instead of retreating, they determined to assert a freedom to deal with such matters, much as they had in 1610. They penned a Protestation, claiming that they possessed, as an "inheritance", the right to freely debate the "arduous and urgent affairs concerning the king, state and defence of the realm" and "bring to conclusion the same".[33] The desire to establish their freedom of speech in these areas was part and parcel of the emergence of the freestanding national interest.

Modern historians have often preferred not to recognize these clear lines of division between king and Commons and have sought to downplay the constitutional significance of the episode. Thus Conrad Russell somehow managed to interpret the Protestation not as a challenge to the role of the crown but rather as a defensive attempt to preserve the privilege of free speech.[34] True, the Commons had long possessed a right of free speech, but it had been restricted to certain permitted areas and never extended to a freedom to discuss whatever they liked, notwithstanding the king's disapproval. Similarly, the claim of an "inherited" right was not a reference to a real past, but the form in which the Commons, since the beginning of the seventeenth century, had begun to create scope for their initiatives by asserting that their liberties were held independently of the crown. James knew that this was a basis on which his overall sovereignty might be infringed, and he reacted accordingly. "We cannot with patience endure our subjects to use such anti-monarchical words concerning their liberties, except that they had subjoined that they were granted unto to them by the grace and favour of our predecessors".[35]

But of even greater radical force was the context in which the Commons were claiming this independent right of free speech. It was not being asserted simply for its own sake. Crucially, it was for the purpose of formulating an independent policy. This was the real, active challenge to the prerogative. The representatives of the political nation were assuming a capacity to determine the appropriate course of foreign policy and to direct it along the lines that they required, irrespective of the fact that it happened to run completely opposite to the intentions of the king. As James put it, "[W]hat have you left un-attempted in the highest point of sovereignty in that petition of yours?"[36]

James could solve his immediate problem by dismissing parliament and leaving himself free to pursue his foreign policy without obstruction. The assembly had demonstrated that he could not expect significant parliamentary aid for anything resembling his own favoured approach. But the Spanish match, with a dowry expected to be in the region of £600,000, might help him to avoid parliaments indefinitely. So, with the Habsburgs in the strongest position they had attained for forty years and the Netherlands again under serious threat, James redoubled his efforts to bring about an alliance with Spain. The absence of parliament at the present made it easier for the king to put the necessary preconditions in place. In the summer of 1622, he made a crucial concession to the Spanish by formally suspending the penal laws against Catholics. This removed what was thought to be the biggest stumbling block to the marriage. It also implemented the aspect of the policy that was most susceptible to political challenge and would have caused a considerable furore had the Houses been in session. Even so, it was not enough to bring the treaty to fruition. The Spanish could never be truly committed to a match with a heretic and were probably happy to regard the talks as a delaying tactic. Negotiations therefore proved inconclusive, and in February 1623, Prince Charles and the Duke of Buckingham set off for Spain in order to try to resolve the matter in person.

The fact that they felt the need to leave the country in comic disguises did not say much for their relationship with the people of England, and the fact that on hearing of Charles's arrival, the Infanta retired into a nunnery never to be seen by her suitor, did not say much for their prospects of success in Spain. Historians have often concluded that the whole initiative was similarly ridiculous. More recently, some have suggested that it is possible to find a degree of political purpose in the journey and to see it as a serious attempt to establish once and for all what conditions, if any, would bring the Spanish to acceptance.[37] Glyn Redworth's characterization of the exploit as foolhardy probably comes closer to capturing the essence of the approach. The journey was a desperate attempt to fulfil an obsessive and misconceived policy. Redworth underlines how far James and Charles were prepared to go to conciliate the Spanish, offering what amounted to complete freedom for the Catholic faith in England and baulking only at the Spanish demand for the conversion of Charles himself.[38] They were flying in the face of the known wishes and strongest feelings of their English subjects.

But in recognizing the imprudent nature of the exercise and the force of James's subjective need of the Infanta's dowry, we should not underestimate their commitment to the diplomatic aim of recovering the Palatinate. There is little evidence that James and Charles might have accepted the marriage treaty without an undertaking on the Palatinate. The early Stuarts may not have had much grasp of a national concept of foreign policy, but they did have the consistent dynastic purpose of restoring the rights of their family in the Palatinate. Their mistake was to suppose that an alliance with Spain was a realistic means to this end. It was never likely that the Spanish would take it upon themselves to arrange the restoration of the Palatinate. In fact, they preferred to treat it as a matter to be resolved between the Stuarts and the Emperor. Brennan Purcell is probably right to suppose that it was when the Spanish first minister Olivares specifically rejected an agreement to restore the Palatinate that Charles decided there was no further point to the negotiations.[39] Overall, the traditional interpretation seems justified: The Spanish were "merely temporising, to forestall any English military action".[40] The French ambassador phrased it more theatrically: The Spanish had been leading James "par le nez".[41]

The futile escapade had one significant effect. Having put his personal prestige on the line, it was a great humiliation for Charles to return home empty-handed. He and Buckingham now had a powerful reason to adopt a more parliamentary view of foreign affairs. It was also perhaps a consolation to Charles that by failing to acquire a Spanish bride, he had quite by accident become a popular hero among the English. His return, on 5 October 1623, was greeted by an unprecedented outpouring of "universal joy . . . even in small towns the festivities were extraordinary".[42] William Laud, who was not a man to overestimate popular initiatives, described it as "the greatest expression of joy by all sorts of people that ever I saw".[43] It was another demonstration of the depth and unanimity of the anti-Spanish perspective among the English people. This was the enemy against which the new national land banded together and measured its position in the world.

Charles and Buckingham now had their own grievance against Spain, and they instigated an attempt to bring royal policy into line with the national view. Ruigh described it nicely as "Buckingham's identification with the will of the nation".[44] The point should be underlined. It corrects the danger of mistaking the main source of the anti-Spanish policy. Because Charles and Buckingham on their return put themselves at the head of an attempt to implement that policy, there has been a tendency to leave the impression that it was their conception.[45] In actuality, their preferred intention, recently redoubled, had been to promote the opposite policy—a Spanish alliance. The true provenance of the anti-Spanish stance must be clearly recognized, since it was the crucial factor in determining the viability, or otherwise, of English foreign policy in the 1620s. The anti-Spanish agenda emerged as the independently conceived priority of the English political nation. It commanded a powerful consensus in the House of Commons right from the beginning of the foreign affairs crisis in 1620 and was maintained in direct defiance of the wishes of the king. The distinctive and defining influence on the course of foreign affairs throughout the decade was that the Commons had a specific view of the policy that the nation required and would pay for nothing else.

It might have been possible to say that the attempted consensus of 1624 was the one and only time that the worldview of the Stuart regime came to coincide with that of the English political nation, but in fact it is doubtful whether Charles and Buckingham ever saw it in a "national" light, and it was not certain that they could persuade James to make the switch in any case. Only James could give a final decision on foreign policy, and only James could authorize the summoning of a parliament, and even the two people to whom he was most ready to listen would not find it easy to persuade the old king to abandon a stance on which he had come to depend.

It would take a strong, united front in the Privy Council to bring James on board. In arranging this, Buckingham had the disadvantage that he was widely hated in rough proportion to the enormous power that he wielded. Yet he succeeded to a significant degree. There is no greater testament to the broad appeal of the project than that it outweighed the general dislike of the Duke. He could be assured of the support of "popular" lords like Southampton, Warwick, Oxford and Saye simply by declaring for war with Spain. With more bitter opponents like Hamilton and Pembroke, Charles's good offices were important in arranging public reconciliation. On 20 December 1623, the Privy Council voted for a parliament, and James succumbed, at least thus far, to the coordinated pressure.

Buckingham could feel certain that a change to the required policy would facilitate an understanding with the influential leaders in the Commons, like Sir Edwin Sandys, Sir Robert Phelips and Sir Edward Coke. With the same end in view, he secured the release of Sir John Eliot from prison, as soon as the summoning of parliament had been agreed. Care was also taken to secure the sitting of that most redoubtable opponent of the Spanish match, Sir Thomas Crew. James's instinct was usually to neutralize such eloquent and strong-willed figures by expedients like giving them offices or overseas missions that prevented their election. But they could all be gathered into a constructive force behind what they believed to be an appropriate foreign policy.

The personnel of this common front were often referred to as "patriots". Tom Cogswell notes this in his analysis of the alliance, yet he does not really elucidate what this use of the term implied. He fails to examine the possibility that it carried the most obvious and principal meaning of the word—that is, to describe a lover or defender of one's country.[46] The general tendency of recent historians has been to avoid acknowledging the emergence of popular sentiment. In the early Stuart period and indeed the Restoration era, "patriot" tended to be applied to opponents of the royal establishment. Historians have turned this round to suggest that the patriots were defenders of traditional liberties against greed and incompetence in government.[47] This is a definition that has eluded the dictionaries and seems to be exclusive to historians with a vested interest in obscuring the true sense. In fact, allegations of corruption were just a by-product of the central charge laid by the patriots, which was the transgression of the common interest as perceived by the people.

At every stage of its development, the term "patriot" has involved the idea of shared allegiance. Before the second half of the sixteenth century, it described a connection that was relatively loose, as in "fellow countryman". It was from the 1570s onward that "patriot" took on its consistent modern meaning of "lover of one's country". In fact, the desire of various peoples to define their interests in opposition to the power of Spain appears to have been an important factor in the change. The collective, consolidated definition of "patriot" emerged in the course of the Dutch struggle for independence. The same development is seen to have occurred in England at much the same time. In 1587, the Earl of Leicester applied it in the context of "such as make show wholly against the king of Spain and to be the only patriots of their country".[48] In 1607, Ben Jonson gave authoritative voice to the new, national sense of the word: "Such as were known patriots, Sound lovers of their country".[49]

So in 1624, the "patriots" were the *patriots*. They were attempting to assert a vital public interest as perceived by the political nation at large, in the face of a king who had no such conception. This involved the revolutionary assumption of a parliamentary right to identify the national priorities without reference to the position of the crown. It was a measure of the radical force of this agenda, as well as of James's staunch conservative reaction, that even with the backing of the Prince and the favourite, the patriot cause never actually succeeded. True to his own lights, James refused to abandon his dynastic and hierarchical preferences and never relinquished his personal sovereignty in the field of foreign affairs. He was obliged to resort to a good deal of evasion and some defiant displays of self-assertion, but in the last analysis, he managed to avoid authorizing a national war against Spain. Buckingham had been pro-Spanish from 1621 to 1623, but in 1624 he seems to have done what he could, even embracing constitutional innovations, to bring the king into line with the parliamentary view. In spite of all, however, he failed.

Since James himself never gave any indication of a real commitment to the patriot cause, "the whole session was fraught with suspicion".[50] The two operative forces in the contest—the king and the representatives of the political nation—were aware that their true positions remained poles apart.

John Chamberlain's comment was most apposite, noting that when they approached the question of negotiating the subsidy, they were "so wary and cautious on all sides as if they were to treat with enemies".[51] This was the plain fact of the matter: James was aligned with the national enemy rather than with the nation itself. He certainly wanted to recover the Palatinate, but he preferred to do it with the help of the Spanish or at least without offending them.

Pressure, and pragmatic necessity obliged James to invite the Commons to give "their advice in matters of greatest import concerning the state and defence of the kingdom".[52] Specifically, he offered the question of the Spanish treaties for their consideration. It was not entirely unhelpful to the king to be able to place responsibility for the annulment of the now unworkable treaties on the shoulders of parliament. And for the rest, James held resolutely to his fallback position, the firm if somewhat last-ditch defence of his personal sovereignty, that he reserved the right to regard parliament's advice as literally just *advice*, which he had no absolute obligation to follow.

The fact remained, however, that the Commons were being allowed to "meddle" in an area that had been declared strictly out of bounds to them three years before. The Venetian ambassador wondered if this was what James really intended. He thought that the king "said more than he meant, as he spoke as if he were carried away . . . the speech shone more by contrast with the one delivered to the last parliament, as then they called felony what is now submitted to the free discussion of the present assembly".[53] Perhaps James conveyed, or failed not to convey, the impression that he was acting against his will. Nevertheless, the concession was bound to appear to the Commons as a vindication of their view of foreign affairs and the justice of their claim to address the matter.

There was no doubt what parliament's advice would be in respect of the Spanish treaties. "Without a dissenting voice, the Commons resolved to recommend the dissolution of the treaties".[54] The more interesting question was whether the Houses, via Buckingham, could prevail on the king to actually reverse the policy and declare war on Spain. James asked for the unprecedented amount of six subsidies, twice what had ever been granted to Elizabeth at one time. In return, he accepted that a forward policy would be put in place. But the best that could be achieved in the way of an agreed set of war aims was four undertakings so vague as to be virtually meaningless. The government was to agree to the defence of England, the securing of Ireland, the assisting of the Low Countries and the setting forth of The Royal Navy. The last was the touchstone. What the Commons meant by "setting forth the royal navy" was the instigation of a sea war against Spain, which to their mind was the natural occupation of the English navy. The king, however, declined to see it in that light.

So while the Commons strove to find ways to tie James to the required policy, James sought to avoid any real commitment, while at the same time inducing the maximum subsidy. One means proposed for achieving the king's compliance was the "appropriation of supply", an arrangement by which the spending of the money would be monitored by parliamentary treasurers. This initiative has been the subject of much debate. Traditionally, it was seen as one of the constitutional innovations through which the Commons were attempting to assert their position. But recent revisionist historians, determined to deny the idea of parliament "advancing", have sought pretexts for diminishing the significance of the move. Thus, although Derek Hirst acknowledges that the accountability to parliamentary treasurers was an innovation, he follows Conrad Russell's diversionary suggestion that this was not constitutional aggression because it came from an agreement between James and Buckingham, to resolve a "procedural" problem as to whether a decision on war should precede supply.[55] In truth, it seems that the idea was first floated among members of the Commons as a condition of their assistance and was taken up by Buckingham to ensure their support. This genesis is evident in the Duke's own memorandum on the matter.[56] And the issue was far from just procedural. At stake was whether the Commons could impose their view on the king or whether James could preserve his right of discretion in foreign affairs. In the event, James accepted the idea only in so far as he could do so without explicitly committing himself to the kind of war that parliament was known to favour.

The revisionist view was so ill-founded that it has not escaped criticism. But counter-revisionist work has tended to focus on merely reasserting that the Commons were attempting constitutional innovations in order to monitor war expenditure.[57] The crucial issue of *why* they were taking these initiatives—that is to say, the strength and substance of parliament's own alternative policy—is somewhat neglected, though this is the vital factor that needs to be recognized. Parliament was attempting to impose its own definition of foreign policy on the king by refusing to pay for anything else. This was the real challenge to the prerogative.

A sea war to undermine the imperial power of Spain was the parliamentary preference, and it reflected the dominant and persistent anti-Spanish sentiment in the country. But forgetting for a moment the basic weakness in government that James created by depriving himself of the support of his people, what were the practical merits of the popular view? Even those who have recognized the force of public opinion have wondered whether it was realistic. Thus, George Trevelyan suggested that the favoured parliamentary policy of a diversionary war at sea and assistance to the Netherlands was naïve and had no more chance of getting the Habsburgs to disgorge the Palatinate than did James's policy of conciliating Spain.

Such criticisms fail to take account of the fact that the recovery of the Palatinate was never parliament's most essential war aim. In 1621 and indeed earlier, MPs would no doubt have supported a serious attempt to defend the Palatinate if it had been proposed at a time when it had some chance of succeeding. But the advocacy of a sea war against Spain was not conceived primarily as a means of achieving that end. War against Spain was always an end in itself. The concept of attacking Spanish naval, financial and colonial power, and as a corollary providing assistance to the Dutch, was seen quite simply as the best way of sustaining and advancing English interests and England's position in the world. It was a positive priority, to be undertaken for its own sake. Sir Edward Coke expressed the sentiment perfectly: "War . . . with Spain is England's best prosperity".[58] This indeed was the galvanizing force of the national project, and the outward vision behind which the national, political land had coalesced. The nation's economic strength was to be gathered from across the oceans, in the founding of a commercial empire, to be forged in the fight against Spanish power.

So the policy put forward in the Commons was part of the continuing pursuit of a clear and powerful vision, not the backward-looking idealization of an Elizabethan "myth". It was a settled agenda, and it was not unrealistic, either in terms of advancing those stated national ambitions, or even of addressing the immediate situation on the continent. Although a sea war against Spain was not proposed merely or even mainly as a means of recovering the Palatinate, it probably had as much chance of achieving that as any other approach. In 1624, MPs recognized specifically that direct military action for that purpose was no longer a serious option. Sir Francis Seymour wanted the enemy and the war aim to be made explicit, taking into account that a Palatinate campaign was beyond their resources and "far from our thoughts". He said that what they could usefully sustain was a naval war against Spain and the provision of aid to the Low Countries.[59] Sir Benjamin Rudyerd recommended the same formula and said that they should be prepared to dig deep to finance assistance to the Dutch and the equipping of the fleet.[60] Thomas Wentworth, Recorder of Oxford, reiterated that it was in England's vital interests to aid the Low Countries. Sir Edward Coke confirmed that that a war at sea, combined with aid to the Low Countries was the most effective policy and the most acceptable to the constituencies. The Palatinate was beyond their reach he said, but not so the Americas. He vowed that he hoped to "live to see the king of Spain lose his Indies".[61] Here again was the beauty of the vision. The Americas were much further away but somehow more accessible. It was indeed, "the most facile and hopeful way for this kingdom to prevail against the Spaniard".[62] "Let us remember," said Sir John Eliot, "that the war with Spain is our Indies, that there we shall fetch wealth and happiness".[63] William Nyell of Dartmouth looked forward to the Spanish navy being "swept from the sea". This sounded somewhat grandiose, but Nyell had specific experience of such an effect. In 1584, English naval action had cleared the Newfoundland waters of Spanish shipping, and the West Country fishing fleets had thrived from then on. This, repeated on a global scale, was the basic idea behind the national enterprise.

Coke also underlined the now settled parliamentary tactic of measuring their intention of supply according to whether they expected to get what they wanted. "If the king will make a war and require no aid he may do it where he will, but if he demand aid he must be advised".[64] This was a new doctrine for a new situation, where the Commons had a definitive idea of what they required. The insistence that supply must be conditional on the adoption of their chosen policy was echoed by a series of prominent speakers, such as Sir Edwin Sandys, Sir Robert Harley, Sir Henry Mildmay, Sir Robert Phelips and Sir Henry Vane, between 11 and 19 March 1624. The force of this position was summed up by Sir John Savile's proviso that allowing parliament's commissioners to account for the money was only to the purpose if it was being spent for the desired ends. The nature of the constitutional aggression was manifest. It lay in the fact that the Commons were seeking to use their power of supply to determine the foreign and military policy of the kingdom, a matter that would once have been entirely within the prerogative of the crown.

The problem was how to tie James down to it. The four points were not sufficiently specific to commit the king to anything. Nor was the "appropriation of supply" adequate to restrain him, for it gave the Commons no control over how a war would actually be run. James duly confirmed their fears, reiterating that the final decision on these questions must remain with him. "Whether I shall send 20,000 or 10,000, whether by sea or land, east or west, by diversion or otherwise by invasion upon the Bavarian or Emperor, you must leave that to your king".[65] In fact, he made it clear that his concern was limited to the Palatinate and that he regarded the Emperor as chiefly to blame for the problem there. He "saw no cause as yet to make the Spaniard the capital enemy".[66] Indeed, at this time he was assuring the Spanish envoys that he had no intention of going to war against Spain and that he still looked to Philip to assist constructively in the restoration of the Palatinate to his son-in-law. A resumption of negotiations with Spain was not out of the question.

James was making quite a good fist of defending his personal sovereignty. He attempted to turn the subsidy bill further in his favour by demanding that it be altered to include a fifth point, stating that the central aim was to be the recovery of the Palatinate. He had consistently held to this position. But the Commons had with equal constancy asserted the opposite, and their response confirmed the strength of their view and the depth of their disagreement with the king. James's request was categorically rejected as "clean contrary to the intent of the House".[67] The bill passed without the inclusion of the king's Palatinate priority. But the practical power remained with James, and when, on 29 May, he indicated his acceptance of the money, he continued to reject the Commons' definition of its purpose. "He insisted particularly and largely upon the recovery of the Palatinate, although that be not specified in the act".[68] The king still had the last word.

By the same token, he sent to Madrid to assure the Spanish that, notwithstanding the aggressive intent of his subjects towards Spain and indeed towards his own powers, his prerogative remained in practice undiminished. "His majesty is bound no further than to advise with them but not to rest upon their advice, except their advice concur with his majesty's wisdom and piety to guide all things according to the greatest reason for the public good".[69] This illustrated the basic dichotomy. To James, parliament's advice had no independent force. The decision was his, and he continued to frustrate the dominant desires of his people. No sea war materialized. The only military action that ensued was an expensive, futile and disastrous gesture towards the Palatinate. Twelve thousand recruits were sent to serve under Count Mansfield. They were appallingly ill-prepared, and something like half their number died of disease. Many of the remainder deserted to the Spanish commander Spinola. This had a peculiar kind of logic, considering the direction of James's own sympathies. Tragically, there was another mission on which the English contingent could have been more suitably employed—that is, in assisting the Dutch in their struggle against the Spanish forces at Breda. Some of the commanders appear to have favoured that campaign. But James insisted firmly and repeatedly that they should not go to the relief of Breda but direct their efforts just at the Palatinate.[70] James never freed himself of a deep respect for Spain and a detestation of the Dutch as republican rebels. He died still fixated on the Palatinate, yet still desiring that the friendship of Spain should be retained.

James had managed to preserve his personal sovereignty and freedom of decision in the field of foreign affairs. Although he had achieved less than nothing abroad, he had, with a display of some resilience, won a battle in defence of his prerogative on the home front. Yet it was a pyrrhic victory. It came at the cost of making it even more difficult for the crown to preserve its sovereignty in future. Parliamentary hopes and pretensions had been encouraged by the patriot coalition, only to be dashed by the king. This gave his son a difficult inheritance. The broken promise not only created a climate of distrust in Charles's early parliaments, it did permanent damage to the reputation of the crown. In 1624, the general will of the kingdom had been effectively channelled into a united front, which had seemed to vindicate the Commons in the justice of their assumed position as representing the national interest. But this unity of purpose was then wilfully subverted by the king, which served to reinforce the apprehension, first registered in Mary's reign, that the policy of the crown could be completely at odds with the vital public interest as perceived by its subjects. Only James had seemed to stand in the way of national self-fulfilment. This must to some degree have fostered the sense that the monarch was no longer the most reliable or natural judge of the public safety, and that parliament might have to extend its claim to that role.

James's bequest of a foreign policy that was essentially alien to his kingdom was a hospital pass that should be more clearly recognized by historians who suggest that Charles was wholly to blame for the breakdown of relations between crown and parliament. But the new king did forego an opportunity to sidestep the problem. One of the anomalies of the beginning of Charles's reign was his failure to build on the alternative approach that he and Buckingham had attempted in 1624. This was despite the fact that in his opening speech he reminded parliament of his part in that process. "I laboured all I could to bring my father to assent thereto".[71] It might have seemed a good moment to revive that cause and to reassemble that coalition now, yet he showed no inclination to seize the chance. "No attempt appears to have been made to re-engage the allies of 1624".[72] The goodwill that had been created was squandered.

Part of the explanation may be that Charles was not so open now to the parliamentary perspective. It is not always sufficiently understood, but he reintroduced the issue of the Palatinate as the central aim of the war. Parliament had, he said, engaged James and himself "in a war for the recovery of the Palatinate".[73] Whether the phrase was being loosely employed or deliberately measured, it was in effect a direct misrepresentation of parliament's stated view. A war centred on the recovery of the Palatinate was precisely what they had *not* engaged for. This was the war aim that the Commons had specifically refused to allow James to insert into the subsidy bill and that they had declared to be "clean contrary to the intent of the House".

Charles had thereby contrived to begin his reign with a definition of foreign policy that sounded like the one favoured by his father and expressly rejected by the House of Commons. What induced him to court trouble in this way? Perhaps he was not fully aware of the weight that these words carried. Perhaps he was just describing a fait accompli, since an abortive move towards the Palatinate was what had actually occurred. No doubt there was an element of dutiful respect for his father, who had set things so specifically on that course and wished them to remain so. It is also possible that coming to the throne had altered Charles's perspective. He had inherited responsibility for his sister, and family honour may have demanded that the war be aimed at recovering her rights. And perhaps now he was king and his dealings with parliament were final, Charles had begun to feel the force of James's view that the Commons could not be allowed to select the national enemy, not if the royal prerogative was to be preserved. It is clear that Charles quickly identified a popular challenge—"an attempt to undermine the foundations of the monarchy".[74] The main precipitant of this was parliament's attempt to dictate the course of foreign policy and enforce the removal of his leading minister.

Conversely, the Commons' readiness to downgrade the issue of the Palatinate underlined the priority that they gave to the anti-Spanish agenda. They would advance substantial supply only if they were getting this precise policy in return. Edward Alford summed up the mood: "We are not engaged to give for the recovery of the Palatinate, for when it was in the act of parliament as first penned, it was struck out by order of the House".[75] Sir Robert Phelips proposed that they should therefore give an interim grant of just two subsidies. He too denied that they were obliged to give more, since their chosen enemy had not been named.[76] It was again Sir Edward Coke who found the most revealing phrase—that the engagement of the House, "was no other but if the king would turn his weapon against the right enemy, we would supply him".[77] Here was the rub—it had to be "the right enemy", as defined by the political nation.

This seemed to become a test of resolve on both sides. Parliament insisted on a declaration against Spain, but Charles refused to make it, despite the fact that a war declared "against the right enemy" would bring forth maximum supply. It should be recalled that whereas parliament's anti-Spanish stance was definitive and final, Charles's grievance against the Spaniards was personal rather than national and might be reconsidered. If Charles had succumbed to the Commons' demand, he would have been subordinating himself to an independent national perspective. Perhaps he had indeed concluded that in the interests of maintaining the prerogatives of a free monarchy, he could not allow parliament to stipulate the enemy.

The symbolic nature of the standoff was underlined by the fact that Charles's refusal to declare the Spaniards as the enemy did not preclude a desire to take some kind of action against them. This was entailed by the new alliance with France. But a combination of the shortage of supply and the inadequacies of government resulted in the fiasco of the attempted assault on Cadiz. Gardiner suggested that, given the appalling state of English provision and morale, it was as well that they never caught up with the treasure convoy from Mexico because "instead of the English fleet taking the galleons, the galleons might well have taken the English fleet".[78]

Ironically, this was not in any case the kind of approach that the parliamentarians regarded as most desirable and profitable. In this sense, it was Charles and Buckingham who were labouring in pursuit of the Elizabethan legend. The parliamentary agenda had moved on. It was still, of course, the same essential project, extending English interests across the oceans, mainly by dint of displacing Spain. But the focus was now less on the defensive aim of weakening Spanish power in the European arena and more on the actual means of building prosperity for the future, especially by undermining the Spanish position in the West Indies. The Caribbean was the most enticing colonial prospect. But the rewards did not come easily, and competition from the more established Spanish enterprise remained a major obstacle. In this respect, the navy could provide vital assistance.

In his essays on policy, Francis Bacon had extolled the great advantages of sea power, "which is one of the principal dowries of this kingdom". It was, he said, the natural strength of a realm bordered by the sea, and it gave them a unique opportunity to fight the kind of war that served the real balance of their interests. "He that commands the sea is at great liberty, and can take as much or as little of the war as he will". Above all, it was the guarantee of the advance of trade and of prosperity to come. "The wealth of both Indies seems in great part but an accessory to the command of the seas".[79]

The Commons and many of their constituents appear to have imbibed this doctrine word for word. It could indeed be taken as articulating the obvious strengths and prospects of the kingdom. The parliamentarians concluded that the war effort should be directed at the Atlantic and the New

World. Their persistent stress on "opening" the Indies was not a generalization but a specific statement of war aims. This was how they wanted the war defined. This was the policy that they believed to be viable and profitable. This was what they were prepared to pay for, in the hope of a good return. But this was also the war that the government declined to pursue, as was duly noted when the full balance sheet was drawn up.

So, with a Commons fixated on a sea war against Spain and a king disinclined to let them set the agenda yet incapable of contriving a means of realizing his own dynastic aims in the Palatinate, Stuart foreign policy staggered on, lurching into wars and campaigns that the kingdom's representatives would neither countenance nor finance. England's relationship with the outside world became dysfunctional. The agreement with France failed to produce the desired action for the recovery of the Palatinate. The only significant result of the alliance may be regarded, without exaggeration, as an actual perversion of the manifest national interest. In the process of agreeing the French treaty, James and Buckingham undertook, in recklessly casual fashion, to give the French king naval assistance against a Huguenot uprising led by the Duke of Soubise. James endorsed this criminally short-sighted policy with his habitual blind hatred of rebellion. "If Soubise or anyone else takes upon himself to commit such follies in your majesties dominions, I will give every kind of assistance against him, in men, in ships, and in any way in my power . . . if those rascally Huguenots mean to make a rebellion, I will go in person to exterminate them".[80] The Huguenots could remain sanguine at the prospect of James coming to exterminate them, but the English navy was a different proposition. One to one, the Rochelais fleet had been more than equal to the French, but with the addition of the English, it was overmatched. A vital element of the defence of La Rochelle was degraded, and the defeat of the Protestant religion in France significantly advanced.[81] This was a specific contradiction of the English national interest, and it would weigh heavily against the crown when the final account was taken.

Charles's approach to foreign affairs during the Personal Rule of the 1630s had continuing echoes of his father's. Charles's actual capabilities in the field were severely limited by the fact that he had determined to govern without parliament and therefore had no money and no military credibility. This did not prevent him from attempting hypothetical exercises in foreign relations. His proposed policies were not in any case such as would have been readily funded by parliament. There was still an obsessive focus on the recovery of the Palatinate. Gardiner noted judiciously that, in the context of the climactic Thirty Years War on the Continent, Charles really had no policy at all. "The one thing for which he cared was the re-establishment of his sister in the Palatinate. His object was purely dynastic. How it would affect Germany, even how it would affect England, were questions he never thought of proposing to himself".[82]

The fact that his approach was merely dynastic meant that he was ready to consider any alliance that might help him recover the Palatinate. But most often his objectives found him proposing treaties (usually in secret) with Spain. There were detailed negotiations to this end in 1631, 1634 and 1636. The proposition was that Charles would assist in the destruction and partition of the United Provinces, while Spain would help in the recovery of the Palatinate.[83] Once again, these designs directly flouted the national interest. England was spared their implementation because Charles's potential allies knew that he had neither the means nor the will to actually do anything. But Spanish troops and supplies were ushered up the channel, exchange facilities were provided for the payment of the Spanish army, and in 1639 Spanish soldiers were allowed to march over English soil in order to enable them to evade the attentions of the Dutch. Meanwhile, the battle for the trade of the West Indies was carried on by independent companies and English sea captains acting on their own initiative and often obliged to sail under foreign flags.

The story of foreign affairs in the 1630s is just as indicative of the Stuart approach as James's obstruction of the patriot coalition in 1624. Notwithstanding their superficial differences at that time, Charles shared his father's essentially dynastic perspective. They really had no concept of a balanced assessment of the objective national interest. This might not have mattered if England had not been emerging from the medieval world and if parliament had not become so convinced of its own capacity to identify the vital priorities of the new nation.

NOTES

1. See Chapters 10 and 11 of this volume and G. Yerby, *People and Parliament*, (Basingstoke 2008), chaps. 3 and 5.
2. C.H. McIlwain, *The Political Works of James VI and I*, (Harvard 1918), p. 314.
3. Sir Thomas Smith, *De Republica Anglorum*, ed. M. Dewar, (Cambridge 1980), p. 85.
4. Ibid., p. 61.
5. T. Cogswell, *The Blessed Revolution*, (Cambridge 1989), p. 16.
6. D.H. Willson, *James VI and I*, (London 1966), p. 411.
7. C.V. Wedgwood, *The Thirty Years War*, (Harmondsworth 1957), p. 99; S.R. Gardiner, *History of England*, (London 1886), III, p. 24.
8. G. Redworth, "Of Pimps And Princes: Unpublished Letters of James I and the Prince of Wales", *Historical Journal 37*, (1994), p. 403.
9. Gritti to the Doge, 6 September 1618, CSP Venetian, XV, p, 306.
10. G. Redworth, "Of Pimps and Princes", *HJ 37*, p. 403.
11. T. Cogswell, *The Blessed Revolution*, p. 19.
12. Gondomar to Philip II, 15 March 1620: S.R. Gardiner, *History of England*, III, p. 338.
13. T. Cogswell, *The Blessed Revolution*, pp. 20–4.
14. Thomas Scot, "Vox Populi", in *Somers' Tracts*, (London 1809), II, pp. 508–24.
15. *Stuart Royal Proclamations*, ed. F. Larkin and P. Hughes, (Oxford 1973), pp. 495–6, 519–21.
16. G. Yerby, *People and Parliament*, chap. 4, and TNA SP14/118, ff. 54–5.

17. R. Zaller, *The Parliament of 1621*, (Berkeley 1971), p. 19.
18. S.R. Gardiner, *History of England*, IV, pp. 25–6.
19. Ibid., p. 126.
20. National Library of Wales, Wyn of Gwydir Ms 9057E/959.
21. *Diary of Walter Yonge*, ed. G. Roberts, (London 1848), Camden Society 41, p. 40.
22. D.H. Willson, *James VI and I*, p. 421.
23. *Commons' Debates 1621*, ed. W. Notestein, et al., (New Haven 1935), III, p. 460.
24. Ibid.
25. *Constitutional Documents of the Puritan Revolution*, ed. S.R. Gardiner, (Oxford 1968), p. 209.
26. D.H. Wilson, *James VI and I*, p. 421.
27. Devon Record Office, Exeter Corporation Latter Book, 60c L206, L210.
28. *Commons Debates 1621*, ed. W. Notestein, VI, p. 199; II, p. 451.
29. Ibid., IV, p. 441; II, p. 452; III, p. 460; IV, p. 200; V, p. 215.
30. Ibid., III, p. 470.
31. C. Russell, *Parliaments and English Politics 1620–1629*, (Oxford 1979), p. 131.
32. S.R. Gardiner, *History of England*, IV, pp. 249, 253, 257–9.
33. Ibid., pp. 261–2; R. Zaller, *The Parliament of 1621*, pp. 182–3.
34. C. Russell, *Parliaments and English Politics*, p. 142.
35. D.H. Willson, *King James VI and I*, p. 423.
36. Ibid., p. 422.
37. T. Cogswell, *The Blessed Revolution*, pp. 60–1; R. Cross, "Pretence and Perception in the Spanish Match", *Journal of Interdisciplinary History 37*, (2007), pp. 566–7.
38. As established by Glyn Redworth in *The Prince and the Infanta: The Cultural Politics of the Spanish Match*, (New Haven 2003).
39. B. Purcell, *The Winter King: Frederick V of the Palatinate and the Coming of the Thirty Years War*, (Farnham 2002), p. 204.
40. D. Hirst, *Authority and Conflict*, (London 1986), p. 133.
41. Tillieres to Puysieux, 12 October 1623, TNA 31/3/57, f. 257.
42. T. Cogswell, *The Blessed Revolution*, p. 9.
43. Diary of William Laud, 6 October 1623, *The Works of William Laud*, ed. W. Scott and J. Bliss, (Oxford 1847–60), III, p. 143.
44. R.E. Ruigh, *The Parliament of 1624*, (Harvard 1971), p. 15.
45. For instance, M. Young, *Charles I*, (London 1997), pp. 37–47.
46. T. Cogswell, *The Blessed Revolution*, p. 84–5.
47. Ibid.
48. Letter of the Earl of Leicester, 15 November 1587, H. Brugmans, *Correspondentie van Robert Dudley*, (Utrecht 1931), III, 281.
49. Ben Jonson, *Volpone*, Act IV, Scene I.
50. R. Ruigh, *The Parliament of 1624*, p. 1.
51. *Letters of John Chamberlain*, ed. N.E. McClure, (Philadelphia 1939), II, p. 548.
52. R. Ruigh, *The Parliament of 1624*, p. 155.
53. Valoresso to the Doge, CSP Venetian, XVIII, pp. 231–4.
54. R. Ruigh, *The Parliament of 1624*, p. 183.
55. D. Hirst, *Authority and Conflict*, p. 125; C. Russell, *Parliaments and English Politics*, p. 177.
56. R. Ruigh, *The Parliament of 1624*, p. 199.
57. For instance, M.B. Young, "Revisionism and the Council of War 1624–6", *Parliamentary History 8*, (1989), p. 1.

58. Sir William Spring, Diary of Proceedings in the House of Commons, February–May 1624, Harvard, Houghton Library, Ms English 980, f. 105; R. Ruigh, *The Parliament of 1624*, p. 205.
59. *CJ*, I, p. 741; Edward Nicholas, Diary of Proceedings in the House of Commons 1624, TNA SP14/166 f. 91v; Diary of John Holles, BL Harleian Ms. 6383, f. 104v; Diary of Sir Walter Earle, BL Add. Ms. 18,597, f. 95v.
60. R. Ruigh, *The Parliament of 1624*, p. 204.
61. Nicholas Diary, f. 97v; Earle Diary, f. 99v; Holles Diary, f. 109.
62. *Constitutional Documents of the Puritan Revolution*, ed. S.R. Gardiner, p. 209.
63. *CJ*, I, p. 740; Spring Diary, 124; Earle Diary, f. 94; Holles Diary, f. 103v.
64. Nicholas Diary, f. 97v; Earle Diary, f. 99v; Holles Diary, f. 109.
65. J. Rushworth, *Historical Collections*, (London 1659–1701), I, p. 140.
66. Diary of Sir Walter Earle, f. 89.
67. Ibid., ff. 184–184v; Dudley Carleton (son) writing to Sir Dudley Carleton (his father), 17 May 1624, TNA SP14/164/81.
68. Sir Simonds D'Ewes, Diary of Proceedings in the House of Commons, February–May 1624, BL Harleian Ms, 159, f. 130; Sir Francis Nethersole to Sir Dudley Carleton, 2 June 1624, TNA SP14/167/10.
69. Sir Edward Conway to Sir William Aston, TNA SP Foreign (Spanish), 94/30, f. 180.
70. S.R. Gardiner, *History of England*, V, pp. 287–90.
71. *Proceedings in Parliament 1625*, ed. M. Jansson and W.B. Bidwell, (Yale 1987), p. 191.
72. C. Thompson, "Court Politics and Parliamentary Conflict in 1625", in *Conflict in Early Stuart England*, ed. R. Cust and A. Hughes, (London 1989), pp. 171–2.
73. *Proceedings in Parliament 1625*, ed. Jansson and Bidwell, pp. 191–3.
74. C. Thompson, "Court Politics and Parliamentary Conflict in 1625", p. 187.
75. *Proceedings in Parliament 1625*, ed. Jansson and Bidwell, p. 407.
76. C. Thompson, "Court Politics and Parliamentary Conflict in 1625", p. 175.
77. *Proceedings in Parliament 1625*, ed. Jansson and Bidwell, p. 403.
78. S.R. Gardiner, *History of England*, VI, p. 20.
79. Francis Bacon, "Of the True Greatness of Kingdoms and Estates", *Bacon's Essays*, (1597–1625), ed. J.M. McNeill, (London 1959), p. 91.
80. S.R. Gardiner, *History of England*, V, pp. 305–6.
81. D. Parker, *The Making of French Absolutism*, (London 1983), p. 54.
82. S.R. Gardiner, *History of England*, VII, p. 169.
83. Ibid., pp. 173–5, 367.

15 "A Declaration of the State of the Kingdom"

The National Imperatives That Necessitated Automatic Parliaments, and the Triumph of Freedom of Trade

The first opportunity that the political nation had to reassert its position was with the calling of the Short Parliament in April 1640. Charles had failed to assemble a viable army to confront the Scottish Covenanters. This humiliating failure had demonstrated his isolation from the constructive force of his English kingdom and the basic inadequacy of his finances. It showed how far he was from encompassing any kind of state interest. There seemed no way that he could gather the resources to sustain a military effort. In the winter of 1639, he was reduced to exacting a forced loan from courtiers and office holders, but it could only be collected from those who were directly dependent on the crown and then only on the assurance that parliament would be called to underwrite it.

Charles's chances of finding compliance in parliament were not increased by the fact that he was seeking support against a Scottish army that he had provoked into existence by attempting to force a strict episcopacy on the Kirk. It was a policy that many of his English subjects also found quite inappropriate. One of the most misleading trends in recent historiography has been the supposition that Charles was addressing a British Problem of differences between his kingdoms.[1] This is a diversion that seeks to avoid acknowledging the actual problem that confronted Charles in England. There was no British problem. Charles was not engaged in trying to forge true integration. His religious preferences were almost as unacceptable to his English subjects as to his Scottish. Nor was he seeking to develop a more coordinated "absolutist" power. Charles had no concept of a "state". One of his real problems was that the idea of the state had been appropriated by the parliamentarians—as a specifically English state. Charles and Laud had come together in 1626 as a front of hierarchical order against an English parliament perceived to be leading a popular challenge to the powers of the crown.[2] Laud's prescription was wholly conservative—restoring the principle of hierarchy in the most explicit Episcopal form. In trying to impose this on Scotland, Charles created a needless problem there and brought to a head the standing problem that he faced in England.

The recent generation of historians, with a preconception against the force of long-term causes, have tended to suppose that parliament assembled in

April 1640 with no particular anti-government agenda, but with a readiness to cooperate, and that it was only their premature dissolution after a few weeks that produced the angry mood of the following assembly later that year.[3] This is to turn the logic and the facts of the situation upside down. The Short Parliament was dissolved precisely because it was *not* ready to cooperate. It is clear that this parliament displayed the same strategic oppositional tendencies that had caused the crises of the 1620s and that furthermore it precisely prefigured the radical initiatives that characterized the opening months of its successor.

Nor had eleven years of consideration, or the current emergency, served to bring the government to a more practical view of the way that parliament would want to do things. The Lord Keeper's opening speech took just the same approach that the Commons had found unacceptable in the 1620s. Immediate supply was demanded, while grievances might be addressed in due course. Except that one important grievance was flouted right at the start. The government had prepared a tonnage and poundage bill in order to try to pressurize parliament into agreeing to grant the levy as from the beginning of Charles's reign. This confirmed what the king's purpose had been in 1629. But it was precisely what parliament had not been prepared to give him unless he relinquished the prerogative of impositions. So they began this assembly at the same impasse that had ended the last.

The elephant in the chamber was the fact that many MPs had more sympathy with the Scots than with the king. In an early speech, Harbottle Grimston made the point as clearly as it was possible to do in the circumstances. He suggested that although the idea of a Scottish invasion was to be deplored, the invasion of the liberties of the subject in England was the more immediate and greater danger. He declared that the remedy of these grievances closer to home should be the priority, and that the perpetrators of the evil should be brought to book.[4] Petitions came in from many of the counties, voicing grievances of every description. At court it was said that this gave comfort to the Scots.[5] No doubt it did.

John Pym stepped forward to draw things together in the first version of what would become the official parliamentarian assessment of national affairs. He rejected the idea that there could be supply before redress of grievances, just as firmly as the king had demanded it. He then enumerated at length the many practices that were considered unacceptable. In religious affairs, despite the general distaste for Laudian high ceremonialism, Pym did not demand a stricter programme of reform in the opposite direction. He emphasized instead the need for latitude. He deplored the "over rigid prosecution" of those who had scruples about engaging in some of the official practices. The need to give some leeway to tender consciences occupied him most. He sought no additional proscription of Catholic tendencies, though he objected to them in so far as they were imposed upon others.[6] The latitudinarian message would have brought comfort to many ears, but it directly contradicted the policies of Laud and Charles, in England and in Scotland.

The list of secular grievances, headed liberty and property, was a long one. But Pym began, as he was to end, by identifying the "grievance of grievances"—the long intermission of parliaments. This framed his argument and was emphasized with intent. The absence of parliaments was the grievance of grievances not only because it was the greatest but also because it was thought to allow the other complaints to fester. The Long Parliament's priority, a few months later, would be a very radical measure to ensure against another such intermission.

The catalogue of specific complaints began with those of greatest significance and longest standing. First came the king's enforcement of impositions, as well as tonnage and poundage without parliamentary consent. These issues remained interdependent, as they had been since 1625. There would be no grant of tonnage and poundage until the power of impositions had been relinquished, and parliament established in the right of setting the customs dues in general. Pym stated this quite plainly as a central parliamentarian purpose, as it had been since 1610. He also condemned the government's revival and extensive use of the practice of monopolies, a grievance of similar longevity, and also part of the core freedom of trade agenda. These long-term grievances are of most importance because they reveal parliament's main, radical ambitions, which were still in place.

Then Pym moved on to the more recent issues, the various fiscal expedients to which the government had resorted during the decade of Personal Rule—the compositions for knighthoods, the extension of ship money, the stretching of the forest laws, the revival of obsolete statutes against nuisances and the recent military levies, like coat and conduct money, without remuneration. None of these things were strictly against the law. In the 1630s, the Caroline government gave every indication of trying to stay within the limits of law and convention. The king's policies were part of an attempt to govern without parliament, but that too was perfectly legal.

Pym returned, though, to the grievance of grievances, the long intermission of parliaments. The stress that he placed upon this gave a fair indication that if the opportunity arose, parliament's priority would be to find some way of ensuring that it could never happen again. The most interesting sign of an intention to enhance the position of parliament was Pym's high estimation of the representative assembly as the seat of "reason" in the kingdom. He suggested that the powers of parliament were to the body politic "as the rational faculties of the soul to a man". This, as Gardiner pointed out, contradicted the traditional notion, which was certainly held by Charles and Laud, that the king himself was the soul of the body politic.[7] In terms of the main theme of the present study, this encapsulated the transition by which parliament was supplanting the king as the essential determinant of the unitary public interest.

It is difficult to understand why some recent historians have supposed that the Commons refused to take their cue from Pym.[8] It seems clear that in general terms they followed the line that he had set out. His keynote speech, restrained yet powerful and persuasive in its detail, "carried the House with it".[9] They were not always patient with long, sober analyses, but they recognized this as a timely survey of the essential public interest. With the stage thus set, the Commons began to give formal shape to their main grievances and prepared to draw up a statement of the case against the crown in respect of ship money and impositions. Convocation and the Lords, under great pressure, voted that the king should be supplied at once, before redress of grievances. The Commons stood firm. On 23 April they resolved that, "till the liberties of the House and the Kingdom are cleared, they know not whether they had anything to give or no".[10] The king then demanded a response to his request for supply. In the face of the monarch's insistence, the Commons conceded that they might defer seeking redress of their grievances regarding the church and parliamentary privilege. But they remained firm on the central grievance of the kingdom regarding the rights of public finance. They were clear that they would give nothing until they had received a concrete endorsement of the principle that money should only be raised only by parliamentary consent.[11] On Strafford's advice, Charles agreed to the idea that the ship money judgement should be taken before the Lord's on a writ of error, when it would undoubtedly be reversed. A deal briefly appeared possible on this basis. The king demanded twelve subsidies. Some MPs were prepared to consider giving six in return for the judgement against ship money. But others declared that their constituents had a greater problem with other charges, like coat and conduct money. The prospect of agreement receded.

Meanwhile, the undercurrent of sympathy for the Scots was coming to the surface. Some of the Commons' leaders had invited their northern neighbours to lay their grievances before the House in a Declaration, scheduled for early May. But with the threat of imminent dissolution in the air, it was agreed that Pym should bring the matter forward and that a petition should be drawn up asking the king to come to an accommodation with the Scots. Charles got word of this proposal and was thus in a position to avoid the final indignity of having to receive such a petition. In Council, Henry Vane gave the opinion that in any case the Commons could never be prevailed upon to give one penny.[12] Parliament was dissolved the following day.

The Short Parliament and especially Pym's guiding speech "laid down the lines upon which, under altered conditions the Long Parliament afterwards moved".[13] This was not surprising, since Pym was setting out a comprehensive assessment of the most vital national affairs. The details did not change, but the overall breadth gradually became apparent. A concern with the kingdom's outward relations was the most obvious sign of the emergence of an independent, unitary public interest, as well as the clearest manifestation of the physical land turning into a coordinated political land. So when Pym sat down to compose the full, completed version of the "declaration of the state of the kingdom", or "Grand Remonstrance" of 1641, foreign policy was not excluded from the catalogue of the governmental sins of Charles's reign. It was high on the list. Pym began by identifying the party that was believed to be subverting the interests of the state, and a principal charge against them was provoking an unwanted conflict with the Scots: "[N]ew canons and a new liturgy were pressed upon them, and when they refused to admit of them, an army was raised to force them to it . . . a most dangerous division and chargeable preparation for war betwixt your kingdoms of England and Scotland".[14] Charles's Scottish adventure was plainly held to be against the best interests of his English kingdom and a gratuitous fermenting of discord between his realms. Pym then looked back into the past, to identify other such transgressions, notably, "the loss of the Rochelle fleet, by the help of our shipping". The English navy, said Pym, had been "set forth and delivered over to the French in opposition to the advice of parliament". This had paved the way for "the loss of all the strength and security of the Protestant religion in France".[15] It was not an overstatement of the case. The episode was a contradiction of the manifest national interest and no doubt served to confirm the parliamentarians in their view that they could offer a sounder judgement in such matters.

They did not, however, regard it as a natural or happy position to actually be at war with France, "the interests and councils of that state being not so contrary to the good of religion and the prosperity of this kingdom as those of Spain".[16] This distinction, plainly stated, illuminates what had been the principal sounding board for the self-defining national interest in foreign policy. Concern was no longer focused on the traditional dynastic tensions with Scotland and France. It was centred on the identification of the particular kingdom whose interests had become comprehensively and uniquely inimical to "the prosperity of this kingdom". Spain presented itself as the conspicuous national enemy. Spain was the kingdom with which war was considered to be profitable and, in a sense, necessary. So it had been in 1588, so it had been in 1624, so it was still in 1641.

It was in this context that in the third item of the Remonstrance, Pym looked back to the moment when the crown and parliament had taken divergent courses in foreign affairs, when in fact the king had failed to follow the line that the Commons had identified as the most propitious way to serve the predominant national interest. Pym specified the betrayal of parliamentary expectations in 1624–1625: "the diverting of his majesty's course of wars from the West Indies, which was the most facile and hopeful way for this kingdom to prevail against the Spaniard, to an expenseful and successless attempt upon Cadiz, which was so ordered as if it had been rather intended to make us weary of war than to prosper by it".[17]

The phrase to "prosper by it" is the most revealing. It illustrates the basic force of the new national concept of foreign affairs, which involved the perception of a kind of war that was a means to profit and prosperity. The definition of a national land was founded largely on the idea that enmity and conflict with Spain was a positive way to mercantile and imperial success, a way, in Sir John Eliot's words, to "fetch wealth and happiness" from across the sea. This was the national project, and it had become the transcendent public priority. It was a national rather than a dynastic perspective, and it exemplified the way that parliament rather than the Stuart crown was assuming responsibility for the state interest.

The shaping of a national perspective in foreign policy was the most obvious reflection of the consolidation of a *political land*. The presentation of a distinctive and characteristic ocean-going face to the world was the graphic illustration of the way that the old context—of the undifferentiated, borderless continuum of the ploughman's lands—was turning into a single, separated national land. But of course Pym's Grand Remonstrance also covered many other topics. This was, after all, the nation's manifesto. It was the clear and complete demonstration of the emergence of an independently definable public interest.

The opening broadsides on foreign policy were followed by the condemnation of the serial neglect and mistreatment of parliament by dissolutions and the imprisonment of MPs and other "patriots" in the 1620s. Then came the substantial core of the grievances—the catalogue of complaints against the attacks on property. At the top of the list were the longest running and most radical of the Commons' ambitions: the various aspects of their desire to bring the setting of all the customs dues under the command of representative consent. At the heart of this and repeated again as they had been so many times since 1610 were the interconnected issues of impositions, tonnage and poundage and the setting of the book of rates—all of which parliament was determined to bring within its own control, now that it had the chance. This was the central aim of the freedom of trade agenda, which was the underlying platform of parliamentarianism.

Then came the well known indictments of the government's bevy of fiscal expedients of the 1630s—such as arbitrary customs dues and books of rates, ship money, forest and knighthood fines, military charges, and the other long-standing grievance of monopolies. Selected for special mention were the monopolies in "soap, salt, wine, leather, sea-coal, and in a manner of all things of most common and necessary use".[18] The protest against monopolies was also part of the drive for freedom of trade, which had become central to the national psyche. Parliament had found an opportunity to legislate against individual monopolies in 1624, but for pragmatic reasons they gave exemption to trading companies, which were often the only means to secure a new market. This offered a loophole, which was exploited by the government in the 1630s, when monopolies were granted under the name of corporations in return for an annual payment to the crown. It became a system of indirect taxation. Both for this reason, and in themselves, monopolies were believed to raise prices and hinder manufactures, as well as transgressing the liberty of the subject in the freedom of his trade. It was the first specific issue addressed by the Long Parliament. As Sir John Culpepper said in the House, monopolies "compriseth many grievances".[19] In the Remonstrance, the protest against monopolies was followed by two further items in the same category, complaining of the breach of the subjects' liberty in the freedom of their trades and in the markets.

Then Pym addressed what we might call the domestic aspect of the national project—the consolidation of landed property. He was unhappy with various aspects of the agrarian policies of the Caroline regime. He claimed, for instance, that, "large quantities of common land and several grounds hath been taken from the subject by colour of the Statute of Improvement, and by abuse of the Commission of Sewers, without their consent and against it".[20] This referred to the king's attempt to commandeer the processes and profits of fen drainage. Pym's main target was the arbitrary money-raising activities of the crown, though probably at Cromwell's instigation the complaint incorporated a defence of the cottagers who were losing their common rights on the Fens. Once the king was gone, however, and the drainage scheme was in the hands of the gentry, Cromwell came to see it as a means of improvement and endorsed action against the protests, in line with the usual position of parliament on the issue.

This was also reflected in Pym's specific view about the question of enclosure, which may be regarded as the other aspect of the consolidation of "the land", and the agrarian dimension of freedom of trade. Enclosure was the process by which the old provisions of scattered strip holdings and open commons were steadily being parcelled up into substantial, individual properties. And in this respect Pym's sympathies naturally lay with the larger landowners and commercializing farmers who were engineering the transformation. He protested against the government's recent attempts to obstruct them. He claimed that the 1630s had witnessed a campaign of prosecutions and fines for the supposedly depopulating effects of enclosure and that this had "drawn many millions out of the subject's purses".[21]

He was not exaggerating overmuch. "A flurry of enclosing activity between 1607–30 was succeeded by a decade in which the Privy Council and the courts were actively deterring it".[22] The government's assault on the process of enclosure began in 1630, when grain shortages were again perceived to have become a problem and instructions were given to the JPs of the Midland counties to reverse all enclosures made in the past five years because they were thought to cause depopulation and dearth. In 1632, 1635 and 1636, Commissions for Depopulation were appointed, and special orders against enclosure were issued to the Justices of Assize. About £50,000 was imposed in fines.[23]

It would be a mistake, however, to regard this as just another example of the fiscality of Caroline government. In this instance, they undoubtedly believed in the righteousness of what they were doing. Transgressors were likely to be summoned before Archbishop Laud in Star Chamber to be lectured on the immorality of their actions.[24] Some of the accused pleaded to be allowed to pay a fine as an alternative to being imprisoned. And in some places, pastures were certainly ploughed up. One Northamptonshire landowner was obliged to plough ninety-five acres.[25] From Nottinghamshire, a farmer wrote to the Privy Council with anxious assurances that he had thrown open all his enclosures, according to their instructions.[26] A Mr Roper, who did not manage to repent in time, was fined over £4,000 and confined to the Fleet for enclosing his land, converting it to pasture and evicting his tenants.[27] Between 1635 and 1638, the Council compiled a list of no less than six hundred offenders, and in 1639 they were still being brought before Star Chamber.

The government's anti-enclosure campaign of the 1630s can be seen as a last-ditch stand in favour of the old "moral" concept of the economy. It may appear futile in retrospect, in view of the underlying trend towards commercialization, but the Caroline attempt to resuscitate the moral dimension illustrates the significance of the change that was taking place. Archbishop Laud injected the traditional policy with the authentic note of spiritual fervour. Like Robert Crowley almost a century before, he condemned as irreligious those who pursued personal gain at the expense of the community. "If any man be so addicted to his private, that he neglect the common, state, he is void of the sense of piety".[28] On this principle Laud professed himself "a great hater of depopulations of any kind, as being one of the greatest mischiefs in this kingdom".[29] True to his lights, he sat on the Commission for Depopulation that was coordinating the attempt to stop the mischief. Unfortunately for Laud, the campaign against enclosure was regarded as a violation of the gentry's now settled assumption that they had an absolute right of property, even if sometimes it had to be claimed out of the common lands. Laud's stand against these practices was duly held against him when he was arraigned before the Long Parliament.

The vehemence of Pym's protest in the Remonstrance was provoked in part by the unexpectedly effective nature of the government's assault on the gentry's agrarian interests. It was a rare occasion when the process of enclosure was actually stopped in its tracks. Apart from the Duke of Somerset's crusade in 1548, this was the only time in one hundred and fifty years when the gentry felt their freedom of action in land encroachment to be under serious threat. It was a severe setback, contradicting the progress made in the 1624 parliament towards normalizing a commercial concept of the land. In the same spirit as that assembly had pursued free trade initiatives, it had also reversed the legislation that had been intended to preserve the older concept of the land as a communal facility. It repealed the long series of anti-enclosure statutes passed between the reigns of Henry VII and Elizabeth I, the enactments that had sought for more than a century to protect the practice of open field husbandry and peasant proprietorship of the commons.[30] No doubt the parliamentarians felt that this should have settled the matter and that henceforth enclosure should proceed unimpeded. But the campaign of the 1630s showed that the authorities still had the will to derail the movement. The direct interference with landed interests added significantly to the sense that some limitation of the royal government might be desirable.

The Levellers took up the struggle against enclosure during the Civil War period, but even they came to the issue rather belatedly. The Leveller movement originated among the small traders and artisans of London, and they had nothing against individual property. Their core grievances, such as the main objection to monopolies and trading restraints, were more in line with the central freedom of trade agenda. Opposition to enclosures appeared as a late addition to their programme following sympathetic contacts with the common soldiers in the New Model Army in 1647. Only the Diggers reflected an inherent hostility to enclosure.[31]

There was, in any case, little chance that parliament would respond to the demand that all enclosures should be thrown open and the land returned to common use. The parliamentarians were driven by the central purpose of removing restrictions on the exercise and extension of property rights. They were unlikely to endorse an agrarian policy that contradicted that aim. In fact, the long era of official attempts at legislation to restrain or reverse the process of enclosure was over. The last bill to be advanced with that purpose was introduced in 1656 by one of Cromwell's Major Generals, Edmund Whalley, who found that enclosure caused problems in the area of Leicestershire under his charge and suggested administrative measures to control it. The proposal received short shrift in the House of Commons, and this has rightly been taken as marking the crucial turning point in the history of the enclosure movement: "when for the first time parliament threw out a bill which proposed to check enclosure . . . the last of the long series that sought to preserve the common fields".[32]

Recorded comments in the House summed up the basis on which the Commons now proposed to give free rein to the process of enclosure. William Lenthall, Master of the Rolls, opined that, "he never liked any bill that touched upon property". Edmund Fowell called it "the most mischievous bill that ever was offered to the House".[33] Enclosure was now fully protected by the newly enshrined, absolute power of property, as embodied in the House of Commons. The bill was rejected without a second reading. Henceforth, the only enclosure legislation would be for the purpose of encouraging and facilitating the practice.

This was an important watershed in the decline of the traditional notion of moral economy. It is true that the process of enclosure still had some way to run and that the land of England would not be completely enclosed until the mid-nineteenth century or later. But much of the southern and central areas had been enclosed by the mid-seventeenth century, and after that in any case enclosure operated in a different climate and was in a sense a different kind of process. It had become normative, both in practice and in principle. By the later seventeenth century, the balance of agrarian society had changed. Where there had once been a majority of independent husbandmen, there was now a majority of landless labourers, employed on the enclosed and consolidated farms.[34] Enclosure was now in a sympathetic relationship with the dominant economic model of unrestricted property rights. It was formally classified as improvement. "Any hindrance to enclosure was allocated to the world of backwardness".[35] After the two decades of the Interregnum, those seeking to enclose their lands from the commons no longer faced the attempted restraints and moral censures that the political and ecclesiastical establishment had trained against them in earlier times. "After mid-century . . . the affirmation of a moral order where economic activities were means to social ends—God's and man's—fell from public view".[36]

Enclosure could now rely on the unequivocal support of the nation's representative assembly. To Archbishop Laud, it had been "one of the greatest mischiefs in this kingdom". But John Pym regarded it as a right of property, and to the revolutionary House of Commons, a proposal to restrain it was "the most mischievous bill". The difference reflected two contrasting worldviews. Laud espoused every aspect of the "conditional" concept of property as outlined by Robert Crowley, who asserted that no one was free to do whatever he liked with his possessions, and everyone was obliged to give the king whatever he demanded, for the king was the sole judge of public necessities. In the same way, just as Laud denied the gentry the right to assume a property in the common lands, so he accorded the king a right to demand supply for what he deemed to be the public safety. The parliamentarians, however, were now inclined to feel that they should advance supply only for matters that they could approve.

A comparison between Laud and Pym is instructive in highlighting the political implications of the transition from the open, communal perspective to the national and individual. In various ways, Laud perpetuated the medieval balance regarding the relative status of the corporeal and the "rational". His Reformist opponents, including Pym, accepted the new dispensation, giving greater independence to intellectual processes and relegating the significance of physical phenomena. They were apt to suggest that Laud's beloved altar rails would be more suitably employed fencing off their gardens. Laud, in turn, rejected the idea that the comprehension of the Bible was the sole necessity for salvation. He aimed to restore the full force of orderly ceremonial and hierarchical structure. His reassertion of the need for bodily worship was explicit. "For my own part, I take myself to worship with body as well as soul whenever I come where God is worshipped".[37] Some of Laud's statements seem to display the medieval tendency to treat the mind as integral to the sphere of the body, rather than according it a separate realm of integrity. His justification for elevating the ceremony of the altar over the lessons of the pulpit was that "[t]he altar is the greatest place of God's residence on earth, for there 'tis, this is my body: but in the other, it is at most, this is my word".[38] Tipping the balance back in favour of bodily devotions went hand in hand with a rejection of the right of private judgement and the insistence that the Head of the Church was the one source of truth. A group of separatists arrested in 1632 were told, "If you have any knowledge of God it hath come through or by us . . . When you have reading, preaching, singing, teaching you are your own ministers . . . The blind leadeth the blind, whereas his Majesty is God's Viceregent in the Church".[39]

He applied the same principle to matters of state. He had no conception of parliament as the seat of collective reason, as Richard Hooker envisaged, or as the "rational soul of the body politic", as Pym declared. Laud had a low opinion of parliament. He seems to have regarded it as just an irritating background noise. His problem was that the political nation had come to see parliament in the same light as Hooker and Pym, as the vital expression of the public interest. It was a major charge against Laud at his trial that he had counselled the king against calling parliament and thus, in Pym's words, contrived to "bereave this kingdom of the legislative power, which can only be used in parliaments . . . and is the only means to restore it from distempers and decays". The loss was heavy indeed, when you believed, like Pym, that the process of parliamentary legislation was "that which makes and constitutes a kingdom".[40] This high estimation of parliament was no doubt one of the things that drove Laud to express the opposite view, denying the indispensability of the legislative process and vowing that he would "make a proclamation as available as an act of parliament".[41]

This dichotomy reflected two contrasting definitions of the concept of the common good. To Pym, the common good was the public interest as determined in parliament, as the representative and coordinating body of a consolidated state. Laud, however, saw no inherent need for parliaments and declined to recognize the force of an independently defined national interest. Yet he clearly did have a strong perception of the common good, as opposed to private ambition. Again, Laud embraced the medieval view. He used the word "state" in its old sense. The "common state" was the common "condition"—the welfare of the common people, whose interests were essentially indistinguishable across the kingdoms of Europe. In that perspective, at once local and open, there was no call for the common good to take on a discrete, bounded national form or for a parliament to articulate central imperatives. To Laud, the unitary common interest was not manifested as a representative national view but by the personal judgements or concerns of the crown. To Pym, however, it was not the king but parliament that was "the rational soul of the body politic".[42] The representative assembly had become the proper judge of national imperatives. By the same token, Pym did not share Laud's inclination to draw a contrast between private ambition and the common good. On the contrary, parliament, as the champion of the coordinated public interest, reflected above all the rights of property.

The repositioning of the common good was associated with an important though little noted change in the scope of the representative concept. Traditionally, it was a matter of local and dispersed constituencies being represented *to* the monarch at the centre of the polity. The primary purpose was for the crown to gain the commitment of every locality collectively to the taxes that were being demanded. The reciprocal right of MPs was to present the particular, pragmatic grievances of their own constituents. This was the concept, we recall, to which James and his ministers attempted to restrict parliament in 1610. They wanted MPs to limit their complaints to specific, practical "inconveniences", while the Commons proposed to attack the royal prerogative of impositions in the name of a general principle of absolute property. As this indicated, the role of representation was acquiring a significant new dimension. And in the course of the early decades of the seventeenth century, parliament began to realize its potential as the representative of a unitary, overriding public interest in every important field, whether it be government finance or foreign affairs.

Historians' assessments of the scope of representation have been somewhat confusing. Derek Hirst noted that the extension of the electorate contributed to the "expansion of the political nation" that the gentry could claim to represent.[43] But he did not deal with the implication of emerging national interests. Barry Coward actually inverted it by supposing that the greatly extended electorate created a wider range of local interests and militated against a united standpoint.[44] Conrad Russell's view was even more contradictory, suggesting that "[t]he one conspicuous success of parliaments during this period was their success in getting themselves recognised as 'the representative of the people' ".[45] Yet he did not ascribe any political significance to this. Russell generally saw representation in a localist perspective. Thus he attributed the problems of the 1620s to loyal obstructivism, whereby MPs were protecting the pockets of their constituents but not challenging the powers of the crown. Russell was content to define these contentions as difficulties that were "reflected in" parliaments rather than as difficulties "with" parliament.[46] He supposed it safe to suggest that their one success was in consolidating their representative role because he believed that this could not amount to a constitutional challenge.

This illustrates one of the structural limitations of revisionist history, which is geared to look no further than the surface of the exercise of power and thus obscures the most obvious mechanism of chronological change: "the trace of what precedes and follows".[47] Russell fails to recognize that the increasing scope of representation could have political implications. The transition by which parliament became the representative of a collected people constituted a change in the balance of public initiative, which might well indicate a need for a redistribution of authority, as the Commons came to reflect issues of national importance that often seemed to be in conflict with the preferences and powers of the crown. It was of critical significance that the representative assembly was replacing the monarch as the main focus of the unitary public interest.

The change in the force of the concept of representation was recognized at the time. Contemporary evidence gives a very definite character to the transition that was taking place and to the general nature of the challenge to the royal prerogative. Commentators understood that parliament was now seen as the representative of the *unitary public interest* in a way that would once have been the preserve of the crown. This development was articulated by the foremost political philosopher of the age, Thomas Hobbes. It is well known that Hobbes identified the new market forces and individualist ethic that he believed lay behind the parliamentary challenge to the crown in the Civil War. Crucially, he also elaborated the way that this translated into a political stance. What he noted was in essence the transition from the "conditional" concept of property to the claim of "absolute" property rights. This emerged in the form of a belief that each individual was "so much master of whatever he possessed that it could not be taken away from him upon any pretence of public safety without his own consent".[48] What is less often realized is that Hobbes also perceived the necessary concomitant—the change in the nature of representation by which parliament itself assumed the capacity to judge the "public safety".

Hobbes believed that the idea of representation had two aspects. One, which he regarded as the proper and original role of parliament, was the presentation of the grievances of the localities *to* the crown at the centre of the polity. The other was the representation of the kingdom as a whole *from* the centre. In Hobbes's view, this latter function had always belonged and should still belong to the king. The problem that he now observed was that this traditional concept, whereby the king represented the unitary public interest, was no longer accepted in the kingdom at large—it was being superseded by another.

> I know not how this so manifest a truth, should be of late so little observed, that in a monarchy, he that had the sovereignty from a descent of six hundred years, was alone called sovereign, had the title of majesty from every one of his subjects . . . was notwithstanding never considered as their representative, the name without contradiction passing for the title of those men, which at his command were sent up by the people to carry their petitions.[49]

Hobbes had identified a crucial change and a substantial underlying motive for the parliamentary challenge to the crown in the 1640s. Over the course of the preceding half-century, the practice of the uniform elected representation of the localities had been consolidated to form a single, public perspective. The king's traditional function of representing the interest of the kingdom as a whole was being usurped by parliament's capacity to reflect the unitary force and emerging priorities of a national political land. It was of great importance that in England the developing state interest had become the preserve of parliament, not the crown.

Parliament's role as the focus of the unitary interest of the kingdom was an obvious general definition of the incentive for providing the assembly with a permanent position and influence at the centre of affairs, and it naturally embraced all the particular public preferences and priorities that regular parliaments would be expected to guarantee. The desire of the political nation to enhance the position of parliament grew in line with the increasingly imperative and coordinated nature of the interests that it reflected, as well as the persistent sense that royal policy tended to contradict rather than advance them. I have described elsewhere how a growing sense of the indispensability of parliament for addressing a range of public interests was leading to the conclusion that the representative body had become essential to the requirements of good government and should therefore be equipped with a constant and reliable place in the polity. Most conspicuously, by the 1620s the Commons were willing and able to identify what they perceived as the one acceptable course of foreign policy. This exemplified the kind of independently defined national interest that they now embodied, and constituted a direct challenge to the traditional role of the crown. A more structured attempt to undermine the prerogative and to replace it with the representative powers of parliament was the consistent ambition from 1610 onward to eliminate the king's discretion of imposed customs dues and to bring all such matters under the aegis of parliamentary consent. A further fundamental necessity for the provision of good government was the regular and guaranteed availability of parliament's legislative service, now seen by the political nation at large as a uniquely effective means of addressing the kingdom's problems, great and small, through the sovereign power of statute. The early Stuart kings seemed not to share this view of statute law as an essential public service and felt no compunction about constantly dissolving parliaments before any legislative programme was completed. The fact that it was not the king but MPs and their constituents who were embracing the systematic provision of national, sovereign law is another aspect of the way that parliament, rather than the crown, was appropriating the state interest.[50]

So parliament became the representative focus of public imperatives, just as parliamentary law came to be regarded as the definitive means of addressing them. Frustration grew when this powerful perception continued to be obstructed or ignored. By 1628, "The often abortion of parliaments" and "The neglect of the counsels of parliament" had risen to the top of the Commons' list of grievances. Conversely, the king now saw parliament as threatening rather than useful. His response, the eleven-year Personal Rule, was an attempt not only to do without parliament but also to demote the significance of the special benefits that it was generally thought to offer. Royal apologists questioned the sole sovereignty of statute law. Archbishop Laud, as already quoted, aimed to "make a proclamation as available as an act of parliament". In the same vein, in glosses on the ship money case, Sir Robert Berkeley declared that, "Rex is Lex . . . a living, a speaking, an acting law", and Sir John Finch asserted that the king had prerogative powers to which, "no acts of parliament make any difference".[51] These statements came back to haunt them when they were arraigned before the Long Parliament. In the House, William Pierrepont condemned Finch for saying "that a hundred acts of parliament could not take it away".[52] Like other parliamentarian leaders, Pierrepont believed that representative statute law was the one "unlimited power". It could no longer be dispensed with.[53]

This illustrates the way that the absence of parliaments served only to harden the dichotomy. It increased the level of discontent at being deprived of essential services and created a determination to ensure that it could not happen again. This intention was signalled in the way that Pym, in April 1640, highlighted the "intermission" of parliaments as "the grievance of grievances", the redress of which would prevent the rest from recurring. The result was the revolutionary measure of the Triennial Act of February 1641, which ended the monarch's prerogative of deciding whether parliament should meet, and arranged for the parliamentary system to call itself into being automatically, if the need arose. This gave parliament an independent and permanent place in political life that it had not previously possessed. It was the first measure to be put in place by the Long Parliament and was the central priority and the vital core of the parliamentarian programme. The Triennial Act changed the balance of the constitution and ensured that the principle and practice of representative rights would now play an unavoidable part in the processes of government. Pym hailed it with the thought that this would guarantee the continuous availability of the legislative process "which can only be used in parliaments" and was "that which makes and constitutes a kingdom. . . . the only way to restore it from distempers and decays".[54] It was also intended to give parliament some permanent leverage on the executive side of government. This was made clear by Lord Digby, in his keynote speech on the third reading. To Digby too "the want of parliaments" was "the efficient cause" of the general problem. The solution was the certainty of regular parliaments. "It is the opportunity of being ill that we must take away . . . In chasing ministers we do but dissipate clouds that may gather again, in passing this bill we shall contribute so much as is in our power to the perpetuation of our sun . . . no state can be confident of a public minister continuing good longer than the rod is over him".[55] The perception of a state interest was instrumental in the desire to give parliament a continuous influence on affairs.

Modern historians have in general sought to downplay or circumvent the radical force of the Triennial Act and to concentrate attention instead on the remedial measures that came later in 1641, outlawing the financial expedients and prerogative courts of administration employed in the 1630s. The purpose of restricting and distorting the focus was to present the parliamentary reforms as merely restorative. But in truth the Triennial Act changed the constitution and was the essential basis of the parliamentarian programme. This was why it was carried through with great urgency and determination, against the expressed opposition of the king. Once the provision of automatic parliaments was in place, the specific remedial measures could be delayed because they had in a sense become a formality. The passing of the Triennial Act was intended to put the representative body in a position to ensure that such things could never happen again. In the Grand Remonstrance of November 1641, Pym noted the particular reforms that parliament had achieved thus far, then hailed especially the Triennial Act, which "well considered may be thought more advantageous than all the former" for it served to "secure a full operation of the present remedy and afford a perpetual spring of remedies for the future".[56]

It has become clear that the readiness of the parliamentarians to alter the constitution had a significant philosophical underpinning. The same definitive "absolute" sovereignty that made the statute an indispensable practical provision also worked to enhance the status of representative law in political theory. Parliamentary leaders like Pym, Digby, Pierrepont and Nathaniel Fiennes could routinely exalt legislative authority as "the greatest power". This gave encouragement and validation to the idea of promoting the legislative assembly to an automatic place in the polity. In this respect, the Triennial Act was the resolution of the grand paradox, or structural anomaly created by the idiosyncratic development of the concept of legislative sovereignty in sixteenth-century England, which made the "absolute" power of legislation dependent on the representative consent of the kingdom. The effect of the Act in resolving this tension was inadvertently predicted in the work of Jean Bodin, the first theorist of sovereignty, who had assumed that in a logical political world, an institution that possessed an essential role in the greatest power of legislation would naturally also hold a permanent position in the structures of government. Writing in the 1570s, he noted that the English parliament did not have a permanent place in government, and he thought this was "sufficient proof" that it did not play a necessary part in the process of lawmaking. His first assumption was correct, but it did not entail the second as he supposed, because the consent of parliament was now the essential, authorizing principle of sovereign law in England. By his mistake, he explicated the change that would be wrought by the Triennial Act, which gave parliament a guaranteed, regular position in the polity, of the sort that Bodin regarded as the necessary accompaniment of legislative sovereignty.[57]

The king deplored the Triennial Act as an "alteration" of the constitution but was obliged to accept it because the Commons made it a strict condition of the supply that he needed to rescue him from an impossible military situation in the north. Thus the radical measure of the Triennial Act was pushed through with single-minded determination by a virtually unanimous House of Commons at the very beginning of the Long Parliament. It was indeed the product of the collective will of the strongly expanding class of gentry that now filled the seats of the enlarged Lower House, just as they filled the proliferating countryseats in the villages. This was the class that over the preceding decades had led and articulated the demand to establish the force of representative rights across a wide range of governmental fields, from finance to foreign policy. In this respect, the Triennial Act, which installed the representative body with an unavoidable place at the centre of affairs, quite specifically linked the "rise of the gentry" to the inception of the English Revolution.

The same can be said of another crucial constitutional change enacted as a priority by the Long Parliament in the early months of 1641. In May, they passed the Tonnage and Poundage Act. The king had sought confirmation of his right of tonnage and poundage throughout the previous parliaments of his reign, but the Commons had not been prepared to accede to it unless he relinquished his right of prerogative impositions. They could now bring this demand to fruition. The normalization of representative rights was at the heart of it, as with the drive for automatic parliaments. The Tonnage and Poundage Act addressed what had been the most consistent and radical parliamentarian ambition for the past three decades—to remove the customs dues from the sphere of the prerogative and bring them under the control of parliamentary consent. The bill asserted that tonnage and poundage and impositions had been taken illegally, that is to say without consent. In truth, impositions had never been illegal; it was simply that parliament wished to make them so. The discretion of setting customs dues by imposition was an ancient right of the crown, confirmed by the Law Lords in 1606 and declared by Sir Francis Bacon in 1610 to be inseparable from the prerogative. In May 1641, the Commons eliminated the king's rights in the matter. They would grant tonnage and poundage only on approval, a few months at a time. They then announced that they were taking command of the customs. "It is hereby declared and enacted . . . that it is the ancient right of the subject of this realm that no subsidy, custom, imposition or other charge whatsoever ought not to be laid or imposed upon any merchants . . . without common consent in parliament".[58] The Commons were usurping power over the customs dues, on the principle of freedom of trade, a concept that had emerged in the mid-sixteenth century and that had been first asserted in the House in 1604 and 1610. In the long campaign for freedom of trade, the gentry had stood shoulder to shoulder with the merchants and the commercializing farmers. The central purpose of representative control of the customs was now being fulfilled. As Gardiner concluded, "By this bill, Charles surrendered forever his claim to levy customs dues of any kind without a parliamentary grant".[59] The king's principal independent source of public finance had been wrested from his hands. As suggested above, the triumph of freedom of trade was also the form in which the principle of absolute property was established. The ending of feudal tenures was not pursued with the same urgency. The Court of Wards was not abolished until 1646, and the system of copyholds survived throughout the revolutionary period. But parliamentary control of the customs was central to the inception of the revolution.

Although freedom of trade had been advanced as a right or principle since 1604 and defined the Commons' most persistent and radical ambition of controlling the customs dues, it would not seem the kind of demand to be expressly stated as the motive for revolutionary change. But in point of fact it did come to be articulated in that form during the Civil War period. In 1642, there appeared a tract that sought to establish the core justification for the campaign in which parliament was engaged—that is, removing the rights of discretionary government. The pamphlet was called *A Brief Discourse on Tyrants and Tyranny*. This would have been a harsh characterization of Charles's essentially lawful Personal Rule. But in fact the pamphlet did not define tyranny in the standard way as lawless autocracy. It was condemned rather as the obstruction of the economic autonomy of others in the freedom of their trades and property. "Tyranny may justly be condemned as the greatest calamity because it is in opposition to the greatest felicity, which lies in liberty and the free disposition of that which God and our industry has made ours".[60] So the principal charge against "tyranny" was that it denied the benefit that was now most aimed for—the right to profit freely by one's goods and property. The political elevation of freedom of trade as the obverse of "tyranny" is a measure of the importance that the concept had acquired in seventeenth century England. This was now the essential condition in which people wanted to carry on their trades and live their lives. The pamphlet echoed the basic rationale of freedom of trade as it had developed since the 1550s: The necessary motivation for a viable and prosperous economic life was the liberty to profit to the full. A king could expect the greatest financial support in a commonwealth governed by consent, where men had "the strong engines of private interest to move them".[61] Under discretionary rule, the king's potential income was not so great, "the uncertainty of keeping hindering for the most part the desire of gain".[62] Here, in a nutshell, were the motives for raising the representative assembly and the representative concept to a position of authority through the Triennial Act and the Tonnage and Poundage Act.

Freedom of trade was accorded the same kind of political and philosophical status in Gerard Winstanley's study of the nature of freedom, written a decade later, in the wake of the Revolution. He noted that, now that the scope for liberty had been created, people were debating what the principal focus of it should be. The freedoms he saw in most general demand were economic. There was the freedom pursued by the gentry and yeomanry to be "landlords of the earth" and to consolidate their property against the commons and the king. But first on the list was the associated demand for freedom of trade. "It lies in the free use of trading, and to have all patents, licenses and restraints removed".[63] There could not have been a more precise description of the concept of freedom of trade that has been identified in these pages, developing from the mid-sixteenth century onward. This was what Winstanley observed as the main, active element of the freedom that the gentry, yeomanry and merchants had sought to establish through the struggles of the 1640s.

These were the very powerful, practical benefits behind the determination to enhance the position of the representative body in the winter of 1640–1641. The Tonnage and Poundage Act effectively completed the constitutional change that the Triennial Act had begun, and these two early measures of the Long Parliament mark the true beginning of the English

Revolution. By these enactments, the king was rendered incapable of ruling independently of parliament. Automatic, regular assembly was intended to ensure that the legislative function was placed at the heart of government and to put parliament in a position to exert a general supervisory influence over the conduct of ministers. It would also enable parliament to exercise the comprehensive control over public finance that it had established under the Tonnage and Poundage Act. These were the specific mechanics of the change in the balance of power that underlay the view expressed by contemporaries such as James Harrington that "the dissolution of this government caused the war, and not the war the dissolution of this government".[64] Charles would certainly need to use force of some kind if he were to recover the rights of a free monarchy that had been so jealously guarded by his father.

NOTES

1. Like many such hares, this was set running by Conrad Russell in "The British Problem and the English Civil War", *History* 72, (1987), pp. 395–415.
2. R. Cust, "Charles I, the Privy Council and the Forced Loan", *Journal of British Studies* 25, (1982), pp. 208–35.
3. For instance, D. Hirst, *Authority and Conflict*, (London 1986), pp. 189–90.
4. J. Rushworth, *Historical Collections*, (London 1659–1701), III, p. 1128.
5. S.R. Gardiner, *History of England*, (London 1886), IX, p. 101.
6. Ibid., pp. 103–4.
7. S.R. Gardiner, *History of England*, IX, pp. 102–4.
8. D. Hirst, *Authority and Conflict*, p. 189.
9. S.R. Gardiner, *History of England*, IX, p. 106.
10. Ibid., p. 108.
11. Ibid., p. 112.
12. S.R. Gardiner, *History of England*, IX, pp. 114–7.
13. Ibid., p. 106.
14. *Constitutional Documents of the Puritan Revolution*, ed. S.R. Gardiner, (Oxford 1968), pp. 204, 216.
15. Ibid., p. 208.
16. Ibid.
17. Ibid.
18. Ibid., p. 212.
19. *Commons Journal*, II, p. 24; J. Rushworth, *Historical Collections*, IV, p. 33.
20. "The Grand Remonstrance", *Constitutional Documents of the Puritan Revolution*, item 32, p. 212.
21. Ibid., item 31, p. 212.
22. S. Hindle, "Caddington Common Enclosure Dispute", *Past and Present 158*, (1998), p. 40.
23. R.H. Tawney, *Religion and the Rise of Capitalism*, (Harmondsworth 1956), pp. 176–7.
24. TNA, SP16/314/29.
25. TNA, SP16/475/72.
26. TNA, SP16/404/142.
27. J. Rushworth, *Historical Collections*, pt. 2, vol. I, p. 268.
28. *The Works of William Laud*, ed. W. Scott and J, Bliss, (Oxford 1847–1860), I, pp. 28–9.

29. Letter to Dr. Gilbert Sheldon, in *The Works of William Laud*, VI, pt. 2, p. 520.
30. *Statutes of the Realm*, vol. IV, pt. 2, p. 1239.
31. For the Diggers, see ahead, pp. 251–4.
32. H. Brailsford, *The Levellers and the English Revolution*, (Nottingham 1983), pp. 426–8.
33. *The Parliamentary Diary of Thomas Burton*, ed. J.T. Rutt, (Colburn 1828), I, pp. 175–6.
34. M. Overton, *The Agricultural Revolution*, p. 179; and see above, p. 80.
35. P. Warde, "The Idea of Improvement", in ed. R. Hoyle, *Custom, Improvement and the Landscape*, (Farnham 2011), p. 142.
36. J. Appleby, *Economic Thought and Ideology in 17th Century England*, (Princeton 1978), p. 70.
37. *The Works of William Laud*, ed. W. Scott and J. Bliss, VII, p. 355.
38. Cited in K. Fincham, *Prelate as Pastor: The Episcopate of James I*, (Oxford 1990), p. 203.
39. S.R. Gardiner, *History of England*, VII, p. 253.
40. J. Rushworth, *Historical Collections*, IV, pp. 201–2.
41. V. Morgan, "Whose Prerogative in Late Sixteenth and Early Seventeenth Century England", in ed. A. Kiralfy, M. Slatter and R. Virgoe, *Customs, Courts and Counsel*, (London 1985), pp. 39–55.
42. S. R Gardiner, *History of England*, IX, p. 102–4; and ibid., p. 154.
43. D. Hirst, *The Representative of the People?* (Cambridge 1975), pp. 192–3.
44. B. Coward, *The Stuart Age*, (London 1994), p. 102.
45. C. Russell, *Parliaments and English Politics 1621–1629*, (Oxford 1979), p. 417.
46. Ibid.
47. Ibid., p. 7.
48. Hobbes, *Behemoth*, ed. F. Tonnies, (London 1889), p. 4.
49. Hobbes, *Leviathan*, ed. M. Oakeshott, (Oxford 1951), p. 122.
50. A full account of the developing polarity between "parliament" and "patent" is in G. Yerby, *People and Parliament: Representative Rights and the English Revolution*, (Basingstoke 2008), chaps. 2–5.
51. S.R. Gardiner, *The History of England*, VIII, pp. 278–80.
52. *The Notebook of Sir John Northcote*, ed. A.H.A. Hamilton, (London 1877), p. 44.
53. J. Rushworth, *Historical Collections*, IV, p. 326.
54. J. Rushworth, *Historical Collections*, IV, pp. 201–2.
55. Ibid., pp. 146–9; for a fuller account of the significance of the Triennial Act, see G. Yerby, *People and Parliament*, chap. 6.
56. *Constitutional Documents of the Puritan Revolution*, ed. S.R. Gardiner, p. 222.
57. Ibid., p. 110.
58. *Statutes of the Realm*, V, p. 104, 16 Car I, cap. 8.
59. S.R. Gardiner, *History of England*, IX, p. 400.
60. *A Brief Discourse on Tyrants and Tyranny*, (November 1642), BL E127 (45), p. 1
61. Ibid., p. 2.
62. Ibid., p. 4.
63. "The Law of Freedom in a Platform", (1652), in *Winstanley: The Law of Freedom and Other Writings*, ed. C. Hill (Cambridge 1983), p. 294.
64. *The Oceana and Other Works of James Harrington*, ed. J. Toland, (Dublin 1737), p. 70.

16 The Commercial Landscape

"How Wide the Limits Stand Between a Splendid and a Happy Land"

At Putney, Henry Ireton highlighted the ultimate force of parliament's new unitary representative function. "We all agree that you should have a representative to govern".[1] The defining proposition of the parliamentarian cause was to give the representative assembly a regular and automatic position at the centre of the polity. It meant altering the constitution to remove the crown's discretion over the calling of parliaments, which was done as the first priority of the Long Parliament, in the Triennial Act of February 1641. This was regarded as a necessity to ensure the provision of good government and the observance of national interests in the future. It was the first radical change pushed through by a unanimous House of Commons, and it retained its place at the core of the proposals put forward on behalf of the army by Henry Ireton at the end of the struggle. It was also a fundamental area of common ground between the generals and the Levellers at Putney.[2]

What they disagreed about at Putney was precisely who should elect the empowered representative. The Levellers wanted to extend the already expanded electorate still further to include the main body of husbandmen and artisans, even if they were not freeholders. The generals wanted to retain the existing freeholder qualification in order to preserve the power of property, but this position also had a national dimension. When Rainborough asked what the ordinary soldiers had been fighting for if not the vote, Ireton suggested that even the disenfranchised had a place in the overall representative system. "They should have the benefit of those laws made by the Representative, yet that they should have the benefit of this Representative. They thought it was better to be concluded by the common consent of those that were fixed men . . . that the will of one man should not be a law, but that the law of this kingdom should be made by a choice of persons to represent, and that choice to be made by the generality of the kingdom. Here was a right that induced men to fight".[3]

Rainborough's famous question is rightly regarded as the first airing of the concept of manhood suffrage. But historians have not fully recognized what this reveals about the contemporary context. It was quite unprecedented that ordinary soldiers could in any respect be thought of as fighting for the vote. It illustrates the remarkable power and broad acceptance attained by the representative principle in early seventeenth-century England. And Ireton's reply shows that even within the parameters of a restricted electorate, the defining shape of the "entire society" as the framework of sovereign representation could be sustained.

Ireton's position has several things to tell us about the dynamics of the nation. Just as he echoed Pym's espousal of the principles of the Triennial Act, he also shared his high evaluation of the legislative function: "the great and supreme power of the commonwealth".[4] And even Ireton's defence of the particular rights of those with property illustrates another facet of the way that the concept of nation was structured. His assertion of the "property interest" is well known, but his exact wording deserves closer attention. He wanted the franchise restricted to "those who have any permanent interest in the kingdom, and taken together do comprehend the whole interest of the kingdom . . . those equals that have the interest of England in them".[5] The concept of a property interest had a national as well as an individual dimension. "I mean by permanent and local that is not anywhere else. As, for instance, he that hath a freehold, and that freehold cannot be removed out of the kingdom".[6] But "if a man be an inhabitant upon a rack rent for a year, for two years, or twenty years, you cannot think that man hath any fixed or permanent interest. That man . . . may have as much interest in another kingdom as here".[7] It is England, as well as the property, that is "not anywhere else". It was easier to conceive of a "fixed" property interest, when it was part of a fixed national body. The completed individual holding, taken en masse, made up the self-possessed land of England.

The process of the consolidation of individual properties had gone hand in hand with the growing sense of England as a discrete, integrated physical and political unit. In the context of the Middle Ages, there had not been the same basis for conceiving of fixed interests. In the old "open" dispensation, everyone, lord or tenant, had held lands on rents or obligations, in widely dispersed forms, and with the loosest of barriers and borders. The claims of the king and nobles had often extended into other realms, and the lands of the church were literally anywhere across the whole of Christendom. But there had since emerged a different context, where the gentry and yeomanry filled out a more clearly defined national land, as an occupying majority of absolute property owners, and by the later decades of the seventeenth century, they held three-quarters of the land of England.

Every development reinforced the reflection between nation and property. The destruction of the independent universal church had made the whole land available for secular appropriation and established the shape of the discrete nation-state. The core dynamic had been the absolute sovereignty of statute law, which drew its authority specifically from the capacity of parliament to reflect the consent of the entire kingdom. The same parties that coordinated the representative system were also the principal beneficiaries of the redistribution of the church lands. Ireton could readily conceive of property interests as representing England because over the preceding century, the class to which he belonged had come to hold the greater part of it. And the same principle of consent that validated the sovereign law of the land also underpinned the rights of possession.

So the new shape of the landscape reflected the national dimension—the emergence of "a land" that was more definitively outlined but that was also, commercially, more outward looking. The microcosm of this balance within the kingdom was the enclosed or engrossed farm, which was structurally self-contained yet also actively connected to the extended marketing systems. The individual property and the autonomous nation-state mirrored one another, as examples of a new conception of the "separated whole". And they shared the same outlook on the world beyond, as a market that was there to serve them.

After the watershed of the revolutionary period, the consolidation and enclosure of individual farms emerged free of the shackles of official condemnation and became accepted as normal practice and the necessary way to improvement. In 1656, Joseph Lee could confidently assert that it was natural for men to seek to maximize the profits of their enterprises in line with their self-interest and that furthermore this served the general good, for the capabilities of the enclosed fields were responsible for the increase in yields.[8] This was no longer seriously contested. The principle of "absolute property" associated with the free commercial use of the land was also given definite form, by the political realization of the freedom of trade agenda. This, as we have seen, was enacted as the second priority measure of the Long Parliament, the Tonnage and Poundage Act of May 1641, which removed the discretionary rights of the crown over the customs dues and deprived the king of the means to raise a significant income independently of parliamentary control. Representative consent now reigned supreme in the field of public finance. The gentry, the yeomanry and the merchants had established absolute command of their property and their trade.

The farm of the Yorkshire yeoman Henry Best, described in 1641, typified the fully commercialized individual enterprise that was becoming normative. His farming practice was based on a detailed analysis of method, identifying the techniques likely to prove most efficient and profitable. He cited Thomas Tusser as a reliable guide in many things. Most notably, Best had a precisely targeted marketing policy. He made a careful study of the optimum times at which to sell each particular type of sheep and at which markets the highest price could be realized. "As for the markets for ewes and lambs, they prove quicker and dearer according as Holderness men come in".[9] He knew that good weather on a Tuesday would encourage the Lincolnshire meal sellers to cross the Ouse for the Wednesday market at Beverley and that in the summer months, Malton market was better for barley than Beverley.[10]

His farm was, naturally, enclosed, for both arable and pasture. He recorded the respective yields from a long series of enclosed arable fields. And according to his figures, the advantages of enclosing pasture were continuing to increase. Until the beginning of the seventeenth century, the general estimates had been that enclosure raised the value of land by about half as much again. Best's account suggested, however, that the increase could be threefold. He noted that in his father's time, before the pasture was enclosed, "it was let to our own tenants or others for 2 shillings a land . . . and lastly for 3 shillings a land, but now being enclosed, they will let for three times as much".[11]

Best was a true individualist, impatient and intolerant of the surviving vestiges of communal farming among the villagers. He was not happy to take part in the traditional practice of contributing to a common stock of hay to feed the flocks in protected quarters, while the snow was on the ground. He noted that, "by this means our hay would have been spent in feeding other men's animals."[12] By the same token, at the beginning of May, when the villagers began to put their cattle and oxen out to graze the open fields, he went to great lengths to ensure that none were turned out onto his own land. This involved constructing special gate blocks and carrying out checks in the middle of the night.[13] This is a good picture of the process by which a consolidated, enclosed individual farm was being separated out from the old communal arrangements. But it also indicates how strong the time-honoured common field system had been. Men like Henry Best inhabited a very different world from that of Piers Ploughman, where it had been an acknowledged Christian duty to work for others as well as for yourself. By the second half of the seventeenth century, mainstream farming had acquired a new structure and a new mindset.

Best employed women with wain rakes to gather the stray corn that would once have been left for the poor to glean.[14] This seems to have become standard practice. The poet John Gerrard noted the same trend, in a passage caricaturing the obsessive force of self-interested, acquisitive farming. Here was the farmer's wife, exerting herself to draw the last ounce of profit from the enterprise and illustrating a transformation in economic ethics.

> At length, oppressed with zeal too great to tell
> At Mammon's shrine the greedy matron fell
> One harvest day, with griping thoughts o'erborne,
> As from the poor she raked the straggled corn,
> Three weighty ears escaped, which when she spied
> In want's pale hand, she sickened, dropped and died![15]

The successful commercial farm employed a substantial number of wage labourers. The arrival of a class of wage labourers was indeed the concomitant of the rise of the consolidated farm. At the beginning of the sixteenth century, the bulk of the rural population was made up of independent smallholders. But by the second half of the seventeenth century, the situation had been reversed, and landless wage labourers had come to comprise the majority of the rural workforce. This underlines the changing balance—as the common field smallholder was gradually squeezed out, and rural society came to comprise the substantial yeoman farmers and the labourers that they employed.

Henry Best recruited outside labour to clip his sheep and another group of labourers to wash them.[16] And when it came to haymaking, he sent to Malton market to hire contingents of "moore-folks" from the Cleveland Hills. There were mowers, gatherers, binders and stookers. The same was done for the harvest in September.[17] These employment patterns were now general and noted, for instance, in the arable regions of Norfolk, Suffolk and Sussex. In Pembrokeshire, it was said that George Owen assembled a labour force of two hundred and forty people to get in his corn harvest.[18] This illustrates the economic substance that a consolidated commercial farm could acquire, as well as the scale of the labouring population that it generated. It becomes easy to understand how wage labourers now composed the bulk of the rural workforce.

Many labourers were unhappy with their reduced circumstances. This is not just a logical assumption; there is good evidence of the resentment of the working class who were on the wrong end of the new dispensations of the rural world. One of the constructive consequences of the disruptions of the Civil War period was the scope and incentive it gave to working people to give vent to their aspirations and their grievances. In normal times, they might have found it much more difficult to make their voices heard. The most striking example, indeed one of the most remarkable movements in English history, was the emergence of the Diggers, which was quite specifically a reaction against the institution of wage labour.

The first appearance of a Digger tendency can be found in a petition from the soldiers of some of the northern regiments of the New Model Army in December 1648.[19] They claimed that their rents had been racked up to a level where their holdings had become uneconomic, "forcing them to be hinds, half-hinds, shepherds and herdsmen". "Hind" was the northern term for labourer. These husbandmen soldiers seem to have objected to any form of employment that could be regarded as wage labour. To reassert their independent status, they asked that their properties be made freehold. They also demanded that the enclosed common lands should be returned to the use of the community. Their case illustrates how every economic circumstance was working against the smallholder and tending to complete the cycle of commercialization.

Also in the winter of 1648, came some pamphlets from Buckinghamshire, launching a socio-economic critique of the concept of wage labour as an exploitative and illegitimate practice. "Light Shining in Buckinghamshire" issued a warning to the rich farmers: "God will visit you for your oppressions. You live on other men's labour, and give them their bran to eat, extorting extreme rents and taxes on your fellow creatures". The sequel, "More Light Shining in Buckinghamshire", drove the message home, castigating the rich man as "the other that sweats not at all, yet makes this man to pay tribute out of his labour, by rates, rents and taxes. It is theft".[20] From another Buckinghamshire town of Iver came exactly the same sentiment: "We cannot enjoy the benefits of our labour ourselves, but for the maintenance of idle persons".[21]

They were arriving at what a later century would call the labour theory of value—the idea that the value of an object derives from the labour that the worker has put into it, and it is wrong for this to be diverted to the profit of another. This was fully articulated in the most famous of Digger broadsides, "The True Levellers' Standard Advanced", published by Gerrard Winstanley, in April 1649, as he and his companions assumed the common ownership of St George's Hill in Walton, Surrey, and began to cultivate it on their own account. "The rich by their covetous wit got the poor to work for small wages, and by their labour got a great increase, for the poor by their labour lifts up tyrants to rule over them".[22] The same refrain is repeated in the "Law of Freedom", Winstanley's ultimate and most extended work of political ideas. "All rich men live at ease, feeding and clothing themselves by the labour of other men, not by their own . . . rich men receive all they have from the labourer's hand".[23]

It need not come as a great surprise that the Diggers hit upon a definition of economic relations that is now associated with a different, industrial age. In fact, they could perceive the elements of the labour theory of value most clearly, in a context where wage labour was becoming the general form of agricultural and indeed manufacturing occupation, and could be contrasted with what was remembered or experienced of the past. They were aware of a previous situation where most people had worked their own smallholdings and retained control of the product of their labour. Now they observed a new balance of things, where a majority worked as hired labourers and struggled to make ends meet, while their employers seemed to profit more. They concluded that a substantial part of the value of the labourers' work was being diverted into the pockets of others, and this was not a rightful process. The Diggers' solution was a concept that precluded the possibility of wage labour—the common lands should be taken into communal ownership.

Winstanley's analysis prefigured the world of wage labour because that world was actually coming into existence. If two-thirds of the population were landless labourers by the 1680s, then by that time England had acquired a rural proletariat. The balance would not alter so dramatically again as it had between the 1530 and the 1680s. In the next hundred fifty-year period, through to the 1840s, the proportion of landless would rise much less, to just three-quarters. One of the most revealing aspects of Winstanley's writing is that it captures the fundamental reversal in agrarian relations that had occurred between 1500 and 1650. To read Winstanley and the Digger tracts is to understand the degree to which the processes of enclosure and engrossing had already undermined the world of the smallholder. To these men, the practice of enclosure defines the cycle of consolidation and deprivation.

The gentry and the freeholders are identified by the fact that their land is enclosed. The force of property "hedges in some, and hedges out others".[24]

The overall effect was that the rights of occupation in the land became much less equal. Many small husbandmen were deprived of an independently worked holding, while a minority class of entrepreneurs consolidated and expanded the scale of their farms and properties, enlarging the proportion of wage labourers in the process. It was a significant change in the balance of society. The husbandmen smallholders, who had comprised the great majority of medieval farmers, had possessed a real degree of independence. Notwithstanding their status as tenants, they had controlled the way they worked their farms and used their product. But now economic independence had ceased to be the situation of the many and had become the preserve of the few.

Less transparent but equally critical was the change that this entailed in the way that people related to the land itself. In every sense, it was being put at a greater distance. This applied even to those who had retained their economic autonomy and proprietorship. The substantial farmers and estate owners now exercised a power over the land that was at once tighter and more controlling, yet also more detached, a process of manipulating the character and condition of their fields and farms to accord with the demands of the market and the development of scientific methods. The less successful farmers, who had become labourers, also now dealt with the land at one remove, as and when required by employers, rather than in continuous, unmediated contact. The old way of relating to the land by direct physical occupation was disappearing.

A core passage in Winstanley provides a sense of the kind of ideal, inseparable closeness to the land that was being lost. He was discussing the nature of freedom, as the main topic of debate in the wake of the Civil War. The basis of political freedom was the triumph of parliament in that struggle. Winstanley was another who could be included in the embrace of Ireton's statement that "we all agree that you should have a representative to govern". He hailed the principle of successive parliaments with much the same fervour as John Pym a decade earlier, for "the parliament is the head of power in a commonwealth, and it is their work to manage public affairs, in times of war and in times of peace".[25] But what were the specific freedoms that parliament was required to guarantee? First on the list was what we have called the freedom of trade agenda. To many people, said Winstanley, freedom "lies in the free use of trading, and to have all patents, licences and restraints removed".[26] Not unconnected with this was the freedom sought by landowners to exercise exclusive control over their property. Winstanley accepted that these were indeed freedoms, but they were only half freedoms, which left the dispossessed labourer unfree.

True, overall freedom—"foundation-freedom"—was free contact with the earth, to work it and enjoy it in a direct and immediate relationship. This was a basic freedom because people drew not just their livings but their existence from the earth. Winstanley justified his concept of the common ownership of the land not by any political or biblical precept but from a simple physical logic. People are entitled to that freedom because the land is the essence of their being—they are actually composed of the earth and the properties associated with it.

> Man is compounded of the four materials of creation, fire, water, earth and air; so is he preserved of the compounded bodies of these four, which are the fruits of the earth, and he cannot live without them . . . surely then, oppressing lords of manors, exacting land-lords and tithe-takers, may as well say their brethren shall not breathe in the air, nor enjoy warmth in their bodies, nor have the moist waters to fall upon them in showers, unless they will pay them rent for it; as to say their brethren shall not work upon the earth unless they will hire that liberty of them.[27]

The land is what they are, and to deny them the freedom of relating to it directly is to deprive them of the truth of their existence. In seeking to restore the independence of the working farmer, Winstanley was identifying the closeness of connection that had been lost, the sense of mutuality with the land, and being part of the earth in an unmediated fashion that had been instinctive to the medieval world and inherent in its farming systems.

By the end of the seventeenth century, the new balance of the landscape was established. The small farm of the husbandman had not entirely vanished, but there was now a majority of wage labourers, and the smallholder no longer defined the agricultural scene, which had come to be dominated by larger farms of a hundred to two hundred acres.[28] "At the beginning of the sixteenth century, Bedfordshire was a land of small freeholders: Only a quarter of the land was in the hands of large manorial lords. Three centuries later, the pattern of ownership had changed dramatically and was concentrated in the hands of a few".[29] Similarly, W. G. Hoskins found that at Wigston, near Leicester by the middle of the eighteenth century, two-thirds of the land was now farmed by the tenants of absentee landlords—large institutions, wool dealers, graziers, butchers and the like.[30]

The most important active agent in the commercial agricultural system was the tenant farmer. "Increasingly the tenant farmer was the human pivot on whom the new rural arrangements turned. He was the symbol of a new society that no longer relied on customary relationships based on the ownership or part ownership of land, but on legal and contractual relationships designed for a world of commerce".[31] To ensure long-term survival under these conditions, the peasant farmer had, in effect, to find a way of clambering aboard the bandwagon. "The peasant who was ingenious enough to drive a shrewd bargain with the squire, and thrifty and enterprising enough to keep himself afloat as a tenant farmer in the new husbandry, might look to a better future".[32]

Enclosure was not an absolute requirement of commercialization. The key to the process was the consolidation of larger farms, and this could continue to take place even where the fields remained technically open. But enclosure now characterized the fabric of the landscape, as it had typified the change. It was the brand of absolute property on the land. The effect was encapsulated by John Clare, as the hedges extended their grip across the moors. He recalled a different time.

> Unbounded freedom ruled the wandering scene,
> No fence of ownership crept in between
> To hide the prospect of the following eye.
> Its only boundary in the circling sky
> . . . these are all destroyed
> And sky-bound mores in mangled garbs are left
> Like mighty giants of their limbs bereft.
> Fence now meets fence in owners little bounds
> Of field and meadow large as garden grounds,
> In little parcels, little minds to please
> With men and flocks imprisoned, ill at ease[33]

The only force capable of reversing the advance of the hedges was something that in one sense, though not all, had gone beyond commercialization. This was the expansive landscaping of the great country house parks. The only place where nothing was allowed to "hide the prospect of the following eye" was in the artificially composed open views of the grounds of the great estates.

As another poet noted:

> And now the manor soars, of costly stone;
> The hedges fall, the woods of ages grown,
> The spacious park its naked visage shows"[34]

The extension of the parks had begun by the early seventeenth century. Already at that time, cottages and even whole villages that "offended" the sensibilities of the local lord or squire were being simply moved out of sight. This happened at Conington, Leighton Bromswold and Melchbourne, in Bedfordshire. Later, in the eighteenth century, it was found expedient to expand these private views even further, in line with the increasing wealth, power and horizons of the country house occupants. The growth of Woburn Park "destroyed some of the fields of Husborne Crawley and half the houses in the village".[35] In this area, "by the end of the eighteenth century there was a significant concentration of large mansions, each within its landscaped park and surrounded by a high wall".[36] There was a continual engrossing of smaller properties by the gentry and yeoman farmers.

The character of the landscape had departed a long way from the medieval context of independent yet communal smallholding. In the new dispensation, a small minority of estate owners and substantial farmers presided over holdings that were at once much more extensive yet wholly individualized. An important aspect too easy to forget is that their perspective, though in one sense completely self-contained, in another sense stretched far away beyond the high walls and hedges of the properties. This can be identified through one of the most famous responses to the changed rural scene.

We know Oliver Goldsmith's heartfelt eulogy to the departing "bold peasantry", as it disappeared under the tide of large farms and estates. The peasant smallholder had belonged to a time

> [w]hen every rood of ground maintained a man.
> For him light labour spread her wholesome store
> Just gave what life required, but gave no more.

But now

> One only master grasps the whole domain
> And half a village stints the smiling plain.
> Ill fares the land, to hastening ills a prey,
> Where wealth accumulates, and men decay.
> . . . times are altered, trade's unfeeling train
> Usurp the land and dispossess the swain.
> Along the lawn, where scattered hamlets rose
> Unwieldy wealth and cumbrous pomp repose.
> . . . the men of wealth and pride
> Take up a space that many more supplied.

The clue to the extra dimension is "trade's unfeeling train". At one level, this reflects the fact that agriculture was now thoroughly commercialized, but at another it tells us that wealth accumulation did not come wholly and perhaps not even mainly from the land. Landowners and tenant farmers not only traded with the world beyond but also invested in the markets that were created. There were vast profits to be made.

And so those helpful winds that had dispersed the great Armada at the very outset of the enterprise now brought home an endless flow of riches.

> Proud swells the tide with loads of freighted ore
> And shouting folly hails them from the shore.
> Hoards, even beyond the miser's wealth abound
> And rich men flock from all the world around.
> Yet count our gains, this wealth is but a name
> That leaves our useful product still the same.

This, perhaps, is Goldsmith's most important insight. He understands the true dimensions of the national project and its capacity to realize itself in places far beyond. He is aware that the consolidation of the land by the power of property within went hand in hand with the foundation of a trading empire without. The exploitation of the land of England was inextricably connected with the exploitation of the globe. This exposes the essential nature of the project as excess—the endless accumulation of wealth, "that leaves our useful product still the same", while it destroys the balance between people, and between them and the land.

This is

> . . . how wide the limits stand
> Between a splendid and a happy land.[37]

NOTES

1. A.S.P. Woodhouse ed., *Puritanism and Liberty*, (London 1938), p. 70.
2. For Ireton's core commitment to the Triennial Act and automatic parliaments, see G. Yerby, *People and Parliament: Representative Rights and the English Revolution*, (Basingstoke 2008), chaps. 8 and 9.
3. A.S.P. Woodhouse ed., *Puritanism and Liberty*, p. 72.
4. *The Representation of the Army*, 14 June 1648, in *Puritanism and Liberty*, ed. Woodhouse, p. 406.
5. Woodhouse ed., *Puritanism and Liberty*, pp. 57, 70.
6. Ibid., p. 57.
7. Ibid., pp. 62–3.
8. Joseph Lee, *A Vindication of a Regulated Enclosure*, (London 1656), pp. 7–9.
9. Henry Best, *Rural Economy in Yorkshire in 1641*, Surtees Society XXXIII, (1857), p. 9.
10. Ibid., p. 100.
11. Ibid., p. 129.
12. Ibid., p. 74.
13. Ibid., p. 118.
14. Ibid., p. 59.
15. John Gerrard, *A Remonstrance*, *Oxford Book of Eighteenth Century Verse*, (Oxford 1984), p. 555.
16. Henry Best, *Rural Economy in Yorkshire*, pp. 18, 21.
17. Ibid., pp. 48, 53, 110, 115.
18. A. Everitt, "Farm Labourers", in *Agrarian History of England and Wales 1500–1640*, ed. J. Thirsk, (Cambridge 1967), p. 434.
19. *The Humble Representation of the Desires of the Officers and Soldiers in the Regiment of Horse for Northumberland*, (5 December 1648), cited in H.N. Brailsford, *The Levellers and the English Revolution*, (Nottingham 1983), p. 447.
20. Ibid., p. 445.
21. "Another Digger Broadside", ed. K. Thomas, *Past and Present* 42, (1969).
22. Gerard Winstanley, "The True Levellers' Standard Advanced", (1649), in *Winstanley: "The Law of Freedom" and other Writings*, ed. C. Hill, (Cambridge 1983), p. 85.
23. Gerard Winstanley, "The Law of Freedom", (1652), in *Winstanley: The Law of Freedom and other Writings*, ed. C. Hill, p. 287.
24. Gerard Winstanley, "A Declaration from the Poor Oppressed People of England", (1649), in *Winstanley: The Law of Freedom and other Writings*, ed. C. Hill, p. 100.

25. Gerard Winstanley, "The Law of Freedom", in *Winstanley: The Law of Freedom and Other Writings*, ed. C. Hill, pp. 344–5.
26. Ibid., p. 294.
27. Ibid., p. 295.
28. C. Wilson, *England's Apprenticeship*, (London 1965), p. 250.
29. P. Bigmore, *The Bedfordshire and Huntingdonshire Landscape*, (London 1979), p. 137.
30. Cited in C. Wilson, *England's Apprenticeship*, p. 250.
31. Ibid., p. 253.
32. Ibid., p. 252.
33. John Clare, "The Mores", in *John Clare: Major Works*, (Oxford 1994), pp. 167–9.
34. John Gerrard, "A Remonstrance", *Oxford Book of Eighteenth Century Verse*, p. 556.
35. P. Bigmore, *The Bedfordshire and Huntingdonshire Landscape*, pp. 143–7.
36. Ibid., p. 148.
37. Oliver Goldsmith, "The Deserted Village", *Oxford Book of Eighteenth Century Verse*, p. 529.

Conclusion

The Limits of the Commercial Land: Is the Environment History?

The preceding pages have identified an environmental aspect in the economic dimension of the English Revolution. In the sixteenth and seventeenth centuries, there was a transformation in the physical environment, in the way that the land was occupied, furnished, used and perceived. It was a broad and fundamental change, of a sort that had not occurred for a millennium and that would not occur again for two centuries to come.

The distinctive character of the medieval period had consisted in a context of open and undifferentiated relationships. That world was open in the communal working habits of the villagers, in the dispersed and interlaced pattern of holdings and in the lighter touch upon the land, just as the agricultural field was open to the reception of spiritual influences. The physical and intellectual spheres worked in a continuum. The realm had not yet become a unitary state but was incorporated into the general body of Christendom. From the mid-sixteenth century, this characteristic context was being supplanted by a different world, which in all these respects was tending to separate things out into more individualistic and closed forms.

The change involved a reversal in the system of beliefs that governed agrarian life. In the older context, economic activities had been subject to moral and political constraints, which sought to sustain the existing balance of welfare. The relationships among the village farmers and between the farmers and the land had been cooperative rather than competitive. The taking of undue individual profits had been disapproved as a breach of the social order and of the God-given purpose of economic affairs, which was to sustain the common good. These principles were associated with a non-interventionist approach to farming. There was little innovation, and a greater reliance on spiritual and communal support systems. In the sixteenth century, these dispensations were being overlaid by a qualitatively different ethic. The more substantial or successful farmers assumed the freedom to maximize their profits through commercialization, extended marketing practices and a more manipulative command over the land. This brought increased productivity and an emphasis on the rights of individual property.

There was a widespread redistribution of land as the dissolution of the monasteries transferred the holdings of the church into secular hands, and the processes of enclosure and engrossing steadily swallowed up the dispersed communal holdings of the open fields. The typical smallholder of medieval times was gradually disappearing from the picture. At the beginning of the sixteenth century, the small, independent husbandmen had comprised the majority of the rural workforce. But by the second half of the seventeenth century, the balance had been reversed, and there was a majority of landless labourers, employed on the enclosed and consolidated farms. The principal agents and beneficiaries of these changes were the gentry and yeomanry. Their success was displayed in the first flowering of secular architectural style. From Elizabethan times, farmhouses were being enlarged and equipped to impressive new standards of comfort and convenience. At another level, the self-possessed symmetry of the new manor houses marked the individualist command and cultural pre-eminence of the gentry.

The classic English rural scene of a squire presiding over every village was taking shape. This was not just a social or local phenomenon. As the yeomanry and gentry proliferated, they were established as power points throughout the countryside and came to form a new kind of administrative network, with a more unified public character. England had never possessed large, independent towns, and now the other iconic medieval edifices of the ecclesiastical houses and the castles were gone. In their place the gentry were physically filling out the land and superseding the old, more open relationships of power and occupation. This was the basis on which the medieval kingdom was turning into a bounded, national-political land.

The gentry and the yeomanry were the mainstays of the system of representation, and parliament's traditional function as the representative of the localities acquired a more consolidated shape as it displayed a growing capacity to reflect the new kind of unitary, overriding common interests that were now emerging. The break with Rome, and the laicisation of the abbey lands created not just an autonomous state and a dominant landowning class but also a series of public priorities and national characteristics. This often focused on rights of consent. The independent judicial status of the old church had been destroyed by the development of an omni-competent, secular lawmaking power, which drew its authority from the capacity of parliament to embody the assent of the entire kingdom. This was the inner dynamic of sovereignty, the force of which was confirmed by its broad reception. Sovereign, statute law, made by king-in-parliament, was embraced by the public at large as the definitive means of addressing their needs and problems, both local and general.

With the disappearance of the universal church, secular forces came to possess "the land" in every sense. The crucial development, distinctive to England, was the assumption of a right to use that property freely. The alienation of demesne land into their hands had left the more substantial, yeomen farmers at liberty to commercialize and profit from the advance of sheep farming and the cloth trade. The process of agrarian capitalism became established through the continuous rise of agricultural prices. There was an extension of regional marketing systems, encouraged by the fact that in England, internal trade was exceptionally free of tolls. These trading freedoms and the special status enjoyed by the concept of consent underlay the growing assertion of a claim to "freedom of trade", or the right to possess and trade goods without being subject to arbitrary restrictions or impositions. In the drive for freedom of trade, the House of Commons launched a persistent assault on the practice of monopolies and most controversially and consistently sought to displace the royal prerogative of imposing customs dues. Parliament aimed to bring all public finance under the control of representative consent. This was the essential basis of an assumption of "absolute property".

The right of consent was the common element in the emergence of a series of perceived English liberties, which operated especially in respect of property, the making of law and the freedom of the Gospel. All these vital interests could be confirmed in antithesis to the power of Spain, which was for good measure the major obstacle to English expansion overseas. Spain could therefore be targeted as the manifest national enemy and identified as such without necessary reference to the interests of the crown, sometimes in open defiance of royal policy. The coherent force of the national land was defined by the perception that conflict with Spain was the way to prosperity. This was the platform for the confirmed national project. The kingdom gathered itself as a secure, self-confident base behind its sea defences and prepared to pursue its economic ambitions far across the oceans.

Parliament's assumption of a right to identify the national enemy and other vital public interests brought it into contention with the policies and prerogatives of the crown at many levels. The growing acceptance of parliament as representing and reflecting a series of unitary public priorities covering every field was an obvious, general motive for the first, revolutionary measure of the Long Parliament—the Triennial Act of February 1641, which changed the constitution by ending the king's discretion over the calling of parliaments and gave the assembly a regular and automatic place at the centre of political life. This would ensure the continuous availability of the legislative function, which the political nation now regarded as the essential basis of good government. It would also enable parliament to exert an ongoing influence over the executive.

The process of establishing parliament with a permanent position of leverage was consolidated by the Tonnage and Poundage Act of May 1641, the second priority measure of the Long Parliament, by which the king "surrendered for ever his claim to levy customs dues of any kind without a parliamentary grant".[1] This brought to fruition the freedom of trade agenda and deprived the crown of its only substantial source of discretionary finance. These early measures, effectively destroying the king's capacity for prerogative government, were of more seminal significance than the Civil War itself.

The war was fought by a king trying to recover his realm for the prerogatives of personal monarchy, against a parliament acting as the government of a different kind of structure—the political nation that it now represented. An important aspect of the strength and motivation of the parliamentary side was that they, rather than the crown, had appropriated the concept of a consolidated state.

These revolutionary events were linked to the changing concept of the land in all its aspects. The principle of consent could be applied in every field. As the independent state, or political nation, developed, parliament became accepted as the embodiment of the unitary public interest, the *central* representative, displacing the king in that role. The parliamentarian movement drew its political validation from the idea that the authority of the overriding sovereign law rested on the assent of the "entire society". This was the inner dynamic of unity, in John Pym's words, "that which makes and constitutes a kingdom". And the same principle of representative consent that bound the nation together under parliamentary statute also underpinned the rights of property. The microcosm of the consolidated nation was the consolidated farm. The individual property and the autonomous nation-state reflected one another as examples of a new conception of the "separated whole". They shared the same outlook on the world beyond, as a market that was there to serve them. So the parliamentarians had a very decided view on the other principal form of land consolidation—they gave powerful support to the process of enclosure. The revolutionary period was a watershed in the history of the enclosure movement. For a hundred and fifty years until 1640, it had faced moral censure and legislative restraint from the political and ecclesiastical authorities. But after the mid-seventeenth century, the official opposition died away. Enclosure now became normative, as part of parliament's central agenda for establishing absolute property and freedom of trade. The commercial land had taken shape.

So how fared the grand, global project around which the political nation had coalesced? In its own terms, it was blessed with great success. The level of achievement vindicated and in a sense explained the strength and assurance of the original vision. Spanish power was finally faced down during the Interregnum and declined sharply thereafter. The Dutch loomed briefly as a rival, but their star was also in the descendent. The United Provinces did not possess the same depth of internal resources, economic or political, as the English state. True, for a while the project laboured in the shadow of monarchs who could not commit to all its aspects. Charles II displayed the same weakness as his grandfather for placing himself in thrall to the leading Catholic monarchy, now the French. But France never posed quite the same problem on the oceans, and the project was never again obstructed from within in the way that James had contrived in the 1620s.

In fact, by the end of the seventeenth century, there was little significant opposition, within or without, to the continued extension of English naval and trading power across the globe. It was still necessary to resist the rise of overweening powers on the Continent and useful to have a national enemy to name as a reverse image. That mantle could be passed seamlessly to whichever of the European powers posed the greatest military threat from across the channel or the greatest theoretical challenge to "English liberty". Such powers could be defied in reasonable security, since the "moat defensive to the house" proved adequate for its appointed task. So for the next quarter of a millennium, the new national land went from strength to strength. Few things have been more triumphant in terms of the realization of a chosen vision than the English—or as it became the British—national project of developing sea power, establishing trade routes and colonies and accumulating riches from across the oceans.

But the geopolitical balance altered radically in the wake of the two world wars of the twentieth century. New and commercially more powerful nations emerged, and the British Empire slipped away. Democratic principles have become more powerful since then, at least in theory, and the trend is to pay lip service to the idea that the acquisition of empire is not a truly defensible ambition. Empires, most notably the American version, now tend to be more subtle and deniable. But the acquisition of wealth, which was the real basis of the British Empire, remains the central and global aspiration. Real, economic democracy, which had in some respects existed on the common lands, shows no signs of recovering. In fact, the force of commercial domination appears greater than ever. Corporate empires reach far and wide across the globe, and money assumes an arbitrary and unaccountable power.

The other obvious source of capital accumulation, associated with the exploitation of imperial resources, was the development of a full-fledged industrial system, which emerged in Britain in the late eighteenth century. This too had its origins in the changes of the sixteenth and seventeenth centuries, when the pattern of a growth economy was established. It was based on an assumption of improvement, through the application of technological advances and new methods. It embraced an approach where the motive force of economic activity was the continuous expansion of productivity and profit. This was fundamentally different from the context of the medieval world, where the guiding principles had been concerned with sustaining the given balance of agrarian society. The long-term consequences of the change were momentous, whether for good or ill. The commercialization of agriculture, which is accurately referred to as agrarian capitalism, brought specialization of product and greatly increased output. It was associated with the extension of regional and international marketing. The exceptional development of commercial farming in England goes far to explain why Britain became the first industrial nation. The spread of agricultural prosperity placed an unusual degree of disposable wealth in the hands of the middling sort of farmer and manufacturer. This, together with the emergence of a rural proletariat of labourers who no longer made their own necessities at home, generated the broad basis of consumer demand that underpinned the rise of factory industry in the nineteenth century. It produced a contradictory mix of enormous wealth for some but degradation and subordination for many.

From the middle of the nineteenth century, the process of industrialization was forging the next great transformation of the physical environment. For the first time, Britain acquired a series of large towns, with the dramatic extension of urban centres to accommodate the burgeoning industrial classes. The proletariat tended to be packed into primitive tenements. A little more space and sympathetic form was given to the cottage-style terraces of the clerical and artisan groups. The principal beneficiaries, the stockbrokers and increasingly affluent middle class, favoured the squares. New forms of transport, the canals and railways, were developed to bind the cities of the industrial nation ever more closely together in the consummated integration of the market. The previous chapter suggested that the only power capable of repelling the advance of agrarian enclosure was the expansive landscaping of the country house parks, but there was now another. The thrust of urbanization reflected a redoubled commercial force that could push back the tide of the enclosed fields. The poet Anna Seward noted that at the hub of Britain's "rage commercial", the rapidly expanding Birmingham

> Commands her aye-accumulating walls
> From month to month to climb the adjacent hills
> Creep on the circling plains, now here, now there
> Divergent—change the hedges, thickets, trees,
> Upturned, disrooted into mortared piles
> The streets elongate, and the statelier square.[2]

Industrialization and urbanization represented another level of pressure on the earth. Yet this was a less seminal change in our relationship with the land than was the transformation in the social balance and ethical framework of economic life that took place during the sixteenth and seventeenth centuries and was reflected in the English Revolution of the 1640s. It was a fundamental reversal of the balance between people and environment when the work of the fields ceased to be guided by moral principles and came to focus simply on material prospects. Economic activities were no longer constrained by the injunction that they should not prove inimical to other interests. Sciences and technologies were developed that facilitated the manipulation and exploitation of the natural world, and. in the absence of a moral perspective, these capacities have been employed blindly to serve the subjective purposes of individual profit and national prestige. The human needs of the industrial workforce only gradually entered into the calculation. The maintenance of the character and requirements of the earth itself never entered into it at all.

The emergence in the 1970s and 1980s of a school of revisionist historians who deprived history of an ethical dimension and ignored the force of economic and environmental connections was in a sense the ultimate product of a commercialized culture that has prevented us from maintaining sympathetic relations with our natural surroundings. This was a generation of academics who embraced the blindness, and were content to suppose that history has no useful lessons to convey. In the light of the denial of economic motivations, it is ironic that the only practical product of revisionist history was the creation of careers and incomes. To say that this offered nothing in the way of constructive meaning is not pejorative, for that was the intention. Revisionists would vehemently deny that their work could have any application to the questions of the present. They proposed that history could be analyzed in static "empirical" form, isolated from the reality of what it had become. But the past can never actually be studied "as it was", and the illusion could be sustained only by ignoring the real signs of change. The most powerful evidences of history were thereby concealed, and the rational sequence of cause and effect was brought to an end. It was no longer possible to gain an understanding of how our society had developed.

It was a route to what Eric Hobsbawm called "the destruction of the past . . . one of the most characteristic and eerie phenomena of the late twentieth century. Most young men and women at the century's end grow up in a sort of permanent present lacking any organic relation to the public past of the times they live in".[3] The only thing to add to that judgement is that the idea of a "permanent present" is probably over-optimistic. In truth, the present may not endure very long in separation from the past, for the stopping of chronological time and the rejection of constructive historical development would seem to entail that there is no future either, and this appears to be the way things are going. The recent academic tendency to deny the connectedness of history and ignore the causative effects of economic forces has deprived us of the natural means of explaining our environmental predicament. The individualist inclinations of modern historians may indeed have contributed more to the problem than to the solution.

It was then, in the sixteenth and seventeenth centuries, that the English people discarded the idea of restraining their economic activities in the interests of others and assumed the freedom and capacity to manipulate the land, and indeed the globe, for their own purposes. This began a process of economic growth that brought enormous wealth for some but generally tended to widen the gap between rich and poor. It also put the human economy in an exploitative relationship with the land and the resources of the earth, and this could not be continued indefinitely without distorting the nature of things and undermining the balance of the environment. At the beginning of the nineteenth century, with the arrival of factory industry, the fateful course of global warming was set. After many centuries of stability, the level of carbon dioxide in the atmosphere began its inexorable rise. It took a sharp upward turn with the second phase of capitalist industry, as it expanded broadly in the wake of the Second World War. The graph has acquired the trajectory of a space rocket, and there is no sign of the human economy returning to earth in a viable state. As I finish writing this, a very unwanted and unfamiliar milestone has been reached. The concentration of carbon dioxide in the atmosphere is now one-third higher than at the beginning of industrialization, and this is a level that the earth has not experienced for more than three million years. We are in unknown and probably unsustainable territory.

Yet there is no prospect of a serious attempt to address the question. Our insistence on a right to live for individual profit makes us incapable of changing our lifestyle to take account of the needs of the natural environment. The general effect of recent historiography has been to exacerbate the problem by denying the significance of the economic field and by discounting the force of its most obvious developments. If we were to take literally the more extreme disconnections of revisionist history, whereby it refuses to acknowledge the rise of sixteenth-century agrarian capitalism and even the reality of the nineteenth-century industrial revolution, we would still be living in the middle ages. In some ways, we might say, if only that were true. And by inversion and another of the unintended consequences of revisionism, we come to a kind of understanding. It may well be that the open, communal world of the husbandman, working his lands with the restraint due to his neighbours and a respect for the land as given, was in the longer-term perspective more viable than the commercial model that displaced it. What would our situation be now if we had managed to avoid the distorted worldview that came to measure everything in terms of the subjective desires of the human individual and the nation-state? There are too many imponderables to attempt a precise calculation of what we might have gained if we had managed to sustain the earlier more balanced relationships to a significant degree. But some things can be said with fair certainty. The economic output of the Western world might not have risen so fast, but it would have been more evenly shared. Our lives would have been more real and less destructive. We would have existed more in concert with the earth and less in opposition to it. We would not have lost our sympathetic awareness of the physical land. And we would not be about to render the natural environment unfit for human habitation.

NOTES

1. S.R. Gardiner, *The History of England*, (London 1884), IX, p. 400.
2. Anna Seward, *From Colebrook Dale*, *Oxford Book of 18th Century Verse*, (Oxford 1984), p. 754.
3. E. Hobsbawm, *Age of Extremes*, (London 1995), p. 3.

Index

For Product Safety Concerns and Information please contact our EU representative GPSR@taylorandfrancis.com
Taylor & Francis Verlag GmbH, Kaufingerstraße 24, 80331 München, Germany

www.ingramcontent.com/pod-product-compliance
Ingram Content Group UK Ltd.
Pitfield, Milton Keynes, MK11 3LW, UK
UKHW022002180926
475701UK00003B/157
9780367263850